# Total Digital Transformation

## Making the Leap to Disruptive Technology in Total Quality Management

Thomas Jones, Lina Khan, and Ahmad Elshennawy

QUALITY PRESS

Milwaukee, Wisconsin

Total Digital Transformation: Making the Leap to Disruptive Technology in Total Quality Management
Thomas Jones, Lina Khan, and Ahmad Elshennawy
American Society for Quality, Quality Press, Milwaukee 53203


Publisher's Cataloging-in-Publication Data

Names: Jones, Thomas, 1990-, author. | Khan, Lina, author. | Elshennawy, Ahmad K, 1952-, author.

Title: Total digital transformation : making the leap to disruptive technology in Total Quality Management / by Thomas Jones, Lina Khan, and Ahmad Elshennawy.

Description: Includes bibliographical references. | Milwaukee, WI: Quality Press, Milwaukee, 2025.

Identifiers: LCCN: 2025939498 | ISBN: 9781636941820 (paperback) | 9781636941837 (PDF) | 9781636941844 (epub)
Subjects: LCSH Total quality management. | Quality control. | Organizational effectiveness. | Disruptive technologies. | BISAC BUSINESS & ECONOMICS / Total Quality Management | BUSINESS & ECONOMICS / Information Management | BUSINESS & ECONOMICS / Development / Sustainable Development | BUSINESS & ECONOMICS / Strategic Planning | TECHNOLOGY & ENGINEERING / Industrial Technology | TECHNOLOGY & ENGINEERING / Quality Control
Classification: LCC HD62.15 .J66 2025 | DDC 658.562--dc23ASQ advances individual, organizational, and community excellence worldwide through learning, quality improvement, and knowledge exchange.

Bookstores, wholesalers, schools, libraries, businesses, and organizations: Quality Press books are available at quantity discounts for bulk purchases for business, trade, or educational uses. For more information, please contact Quality Press at 800-248-1946 or books@asq.org.

To place orders or browse the selection of all Quality Press titles, visit our website at: http://www.asq.org/quality-press.

30 29 28 27 26 25 LS 7 6 5 4 3 2 1

QUALITY PRESS
Quality Press
600 N. Plankinton Ave.
Milwaukee, WI 53203-2914
Email: books@asq.org
Excellence Through Quality™

# Contents

*List of Figures and Tables* ........ vii
*Contributors' Acknowledgments* ........ ix
*List of Acronyms* ........ xi

1 WHAT IS QUALITY 4.0? ........ 1
Quality Throughout the Ages: A Brief History ........ 1
A Closer Look at the Quality Stages ........ 6
Five Key Steps for Implementing Quality 4.0 ........ 15
Moving Forward with Quality 4.0 ........ 17

2 WHY QUALITY 4.0? ........ 18
Components of Quality 4.0 Framework ........ 19
Challenges of Quality 4.0 ........ 24
Technologies Driving Quality 4.0 ........ 26
How Ready Are You? ........ 32

3 SYSTEM REQUIREMENTS FOR QUALITY 4.0 ........ 37
Statistical Process Control ........ 37
Six Sigma ........ 39
Quality Data Usage ........ 41
What Makes It 4.0? ........ 42
Automation is a Quality 4.0 Requirement ........ 46
Quality 4.0 System Deployment ........ 52

4 MAKING A DATA STRATEGY ........ 55
Ford Motor Company ........ 56
Toyota ........ 56
Walgreens ........ 57

The Value of Information . . . . . . . . . . . . . . . . . . . . . . . . . . . . 58
Decision Complexity. . . . . . . . . . . . . . . . . . . . . . . . . . . . . . . . 61
Decision Speed . . . . . . . . . . . . . . . . . . . . . . . . . . . . . . . . . . . 61
Decision Point. . . . . . . . . . . . . . . . . . . . . . . . . . . . . . . . . . . . 62
Define the Information. . . . . . . . . . . . . . . . . . . . . . . . . . . . . . 62
The Four Vs of Data. . . . . . . . . . . . . . . . . . . . . . . . . . . . . . . . 65
Example of a Data Strategy . . . . . . . . . . . . . . . . . . . . . . . . . . 70

5 ESTABLISHING SYSTEM PARAMETERS . . . . 77

Data Management . . . . . . . . . . . . . . . . . . . . . . . . . . . . . . . . . 79
Data Collection. . . . . . . . . . . . . . . . . . . . . . . . . . . . . . . . . . . 79
Data Movement . . . . . . . . . . . . . . . . . . . . . . . . . . . . . . . . . . 80
Information Delivery . . . . . . . . . . . . . . . . . . . . . . . . . . . . . . . 83
The Right System Parameters. . . . . . . . . . . . . . . . . . . . . . . . . 86
Cybersecurity Parameters . . . . . . . . . . . . . . . . . . . . . . . . . . . 86

6 COMMON BARRIERS . . . . . . . . . . . . . . . . . . . 88

Acceptance Barriers. . . . . . . . . . . . . . . . . . . . . . . . . . . . . . . . 88
Technology Barriers. . . . . . . . . . . . . . . . . . . . . . . . . . . . . . . . 89
Strategic Barriers. . . . . . . . . . . . . . . . . . . . . . . . . . . . . . . . . . 91
Data and Privacy Barriers. . . . . . . . . . . . . . . . . . . . . . . . . . . . 92
Standards Governance Barriers. . . . . . . . . . . . . . . . . . . . . . . . 93

7 IMPLEMENTATION PLAN . . . . . . . . . . . . . . . . 95

Define Implementation Success and Outcomes . . . . . . . . . . . . . 96
Research and Readiness Assessments . . . . . . . . . . . . . . . . . . . 97
Understand the Goals and Barriers. . . . . . . . . . . . . . . . . . . . . 98
Conduct Resource Planning . . . . . . . . . . . . . . . . . . . . . . . . . 100
Perform Risk Assessments and Risk Planning . . . . . . . . . . . . 102
Establish an Implementation Timeline and Assign Responsibilities . . . . . . . . . . . . . . . . . . . . . . . . . . . . 105
Summary . . . . . . . . . . . . . . . . . . . . . . . . . . . . . . . . . . . . . . . 105

8 REALIZING VALUE FROM QUALITY 4.0. . . 107
Systematically Generating Better Information . . . . . . . . . . . . . 108

9 PUTTING QUALITY 4.0 INTO PRACTICE . 126
The Critical Success Factors (CSFs) . . . . . . . . . . . . . . . . . . . . . . 127
The Model for Implementation. . . . . . . . . . . . . . . . . . . . . . . . . . 129
Validating the Approach . . . . . . . . . . . . . . . . . . . . . . . . . . . . . 131
Walkthrough Example . . . . . . . . . . . . . . . . . . . . . . . . . . . . . . . 132
Technology Strategy Development . . . . . . . . . . . . . . . . . . . . . . 136
Organizational Pre-Work . . . . . . . . . . . . . . . . . . . . . . . . . . . . . 139
Software Down Selection . . . . . . . . . . . . . . . . . . . . . . . . . . . . . 146
Total Quality Management and Lean Deployment . . . . . . . . . 148
Lean and TQM CSFs. . . . . . . . . . . . . . . . . . . . . . . . . . . . . . . . . 151
Post Strategy Phase . . . . . . . . . . . . . . . . . . . . . . . . . . . . . . . . . 153
Other Real-World Examples . . . . . . . . . . . . . . . . . . . . . . . . . . . 156
How Generative AI Helps to Improve Quality. . . . . . . . . . . . . 161
Success Stories from Implementing ERP Systems. . . . . . . . . . . 163

10 THE FUTURE OF QUALITY 4.0. . . . . . . . . . . . 166
Industrial Revolutions: Changing How We Operate . . . . . . . . 167
Implementing Technology . . . . . . . . . . . . . . . . . . . . . . . . . . . . . 168
Defining the Future State. . . . . . . . . . . . . . . . . . . . . . . . . . . . . . 168
Core Competencies: The Value Proposition . . . . . . . . . . . . . . . 169
Quality Cultures and Quality 4.0. . . . . . . . . . . . . . . . . . . . . . . . 171
Total Quality . . . . . . . . . . . . . . . . . . . . . . . . . . . . . . . . . . . . . . . 174
Lean . . . . . . . . . . . . . . . . . . . . . . . . . . . . . . . . . . . . . . . . . . . . . 175
Realizing Long-Term Value from Quality 4.0 . . . . . . . . . . . . . 177
A Thought Example: The Perfect Cup of Coffee . . . . . . . . . . . 178

11 TRANSITIONING TO QUALITY 5.0 . . . . . . . 182
Feeding into Quality 5.0. . . . . . . . . . . . . . . . . . . . . . . . . . . . . . 182
Implementing Quality 5.0 . . . . . . . . . . . . . . . . . . . . . . . . . . . . . 184

Quality for Humanity. . . . . . . . . . . . . . . . . . . . . . . . . . . . . . . 185
Quality 5.0 Technologies . . . . . . . . . . . . . . . . . . . . . . . . . . . . . 188
Benefits, Challenges, and CSFs. . . . . . . . . . . . . . . . . . . . . . . . . . 188
More on Quality 4.0 and Transitioning to Quality 5.0 . . . . . . 191
Society 5.0: The Road to a Smarter Society . . . . . . . . . . . . . . . 192
Industry 5.0 and Society 5.0. . . . . . . . . . . . . . . . . . . . . . . . . . . 196
Society 5.0 and Quality of Life. . . . . . . . . . . . . . . . . . . . . . . . . 197
Looking to the Future. . . . . . . . . . . . . . . . . . . . . . . . . . . . . . . 198

*References* . . . . . . . . . . . . . . . . . . . . . . . . . . . . . . . . . . . . . . . . 199
*About the Authors* . . . . . . . . . . . . . . . . . . . . . . . . . . . . . . . . . 208

# List of Figures and Tables

Figure 1.1: History of modern quality. . . . . . . . 7
Figure 1.2: Quality 4.0 implementation steps . . . . . . . . 15
Figure 2.1: What is Quality 4.0?. . . . . . . . 18
Figure 2.2: Axes of Quality 4.0. . . . . . . . 19
Figure 2.3: Summary of the benefits of Quality 4.0 implementation. . . . . . . . 31
Table 3.1: Quality 4.0 technology elements. . . . . . . . 43
Table 3.2: Example technologies by category . . . . . . . . 51
Figure 3.1: Conceptual model for implementation of Quality 4.0. . . . . . . . 53
Figure 4.1: Technology strategy deployment section of conceptual model . . . . . . . . 58
Figure 5.1: Technology methodology and enablers section of conceptual model . . . . . . . . 77
Table 8.1: System performance . . . . . . . . 115
Table 8.2: Human inspection performance. . . . . . . . 116
Table 8.3: Vision system performance. . . . . . . . 116
Table 8.4: Comparison of human and vision systems . . . . . . . . 116
Table 8.5: Defective parts processed during the inspection period . . . . . . . . 118
Figure 9.1: Framework for critical success factors in Industry 4.0 implementation . . . . . . . . 126
Table 9.1: Critical success factors for implementation . . . . . . . . 127
Figure 9.2: Successful implementation of Industry 4.0 . . . . . . . . 129
Figure 9.3: Quality 4.0 implementation model . . . . . . . . 130
Figure 9.4: Thematic analysis approach. . . . . . . . 131

Figure 9.5: Strategy phase . . . . . . . . . . . . . . . . . . . . . . . . . . . . . 134
Table 9.2: Strategy phase CSFs . . . . . . . . . . . . . . . . . . . . . . . . . . . 141
Table 9.3: CSFs revisited . . . . . . . . . . . . . . . . . . . . . . . . . . . . . . 151
Figure 9.6: Quality 4.0 implementation model . . . . . . . . . . . . . . 153
Figure 9.7: Quality spending strategies . . . . . . . . . . . . . . . . . . . . . 155
Figure 9.8: Quality 4.0 deployment completion . . . . . . . . . . . . . . 156
Figure 9.9: Maturity level of Quality 4.0 in each HEI in Saudi Arabia . . . . . . . . . . . . . . . . . . . . . . . . . . . 158
Figure 10.1: Conceptual model for implementation . . . . . . . . . . 172
Figure 11.1: Characteristics of Quality 5.0 . . . . . . . . . . . . . . . . . . 187
Figure 11.2: Organizations and digital transformation . . . . . . . . 192
Figure 11.3: Economic and social innovation by deepening of Society 5.0 . . . . . . . . . . . . . . . . . . . . . . . . . . . . . . 193

# CONTRIBUTORS' ACKNOWLEDGMENTS

THE AMERICAN SOCIETY for Quality and Quality Press would like to thank the Quality Press Peer Review Committee for its invaluable volunteer participation and contributions to this work. Without our volunteers' subject matter expertise, time, and passion for creating content, none of our efforts would be possible.

## QUALITY PRESS PEER REVIEW COMMITTEE MEMBERS

Scott A. Laman, Chair

Peter Pylipow, Vice Chair

Melvin Alexander

Lance Coleman

Ahmad Elshennawy

Marc Hamilton

Gary Jing

Trevor Jordan

Jane Keathley

Betsy Macht

Mary McShane-Vaughn

Jayet Moon

G.S. Sureshchandar

Tiea Theurer

# List of Acronyms

3D—Three Dimensional

5G—Fifth Generation

5S—Sort, Set in order, Shine, Standardize, and Sustain

AI—Artificial Intelligence

ANSI—American National Standards Institute

AOQL—Average Outgoing Quality Limit

APT—Annual Production Throughput

AR—Augmented Reality

ASQ—American Society for Quality

ASQC—American Society for Quality Control

B2B—Business to Business

BCG—Boston Consulting Group

BPM—Business Process Management

CI—Continuous Improvement

CNC—Computer Numerical Control

CPD—Cost Per Defect

CSF—Critical Success Factor

CSI—Cost of Slow Information

DARPA—Defense Advanced Research Projects Agency

DGQ—German Association for Quality

DMAIC—Define-Measure-Analyze-Improve-Control

DoD—Department of Defense

DOWNTIME—Defects, Overproduction, Waiting, Not utilizing talent, Transportation, Inventory, Motion, Extra processes

DPMO—Defects Per Million Opportunities

DR—Defective Rate

EQMS—Electronic Quality Management System

ERP—Enterprise Resource Planning

FPY—First Pass Yield

GPS—Global Positioning Satellite

HEI—Higher Education Institution

IIoT—Industrial Internet of Things

IoT—Internet of Things

IPv—Internet Protocol version

IPv6—Internet Protocol version 6

IT—Information Technology

KPI—Key Performance Indicators

LPD—Lost Production per Defect

LTPD—Lot Tolerance Percent Defective

ML—Machine Learning

MRP—Material Resource Planning

OC—Operating Characteristic

OT—Operational Technology

PDCA—Plan, Do, Check, Act

PLA—Production Lost Annually

PLC—Programmable Logic Controller

PLM—Product Lifecycle Management

QC—Quality Control

QM—Quality Management

QMS—Quality Management System

R&D—Research and Development

ROI—Return on Investment

SaaS—Software as a Service

SPC—Statistical Process Control

SOP—Standard Operating Procedure

SRC—Scrap/Rework Cost per Part

TAM—Technology Adoption Model

TD—Total Defects

TOE—Technology, Organization, Environment

TQM—Total Quality Management

UTAUT—Unified Theory of Acceptance and Use of Technology

VR—Virtual Reality

WIP—Work in Progress

# Chapter 1
## What is Quality 4.0?

QUALITY IS A term that takes on different meanings, ranging from the totality of characteristics to satisfying a given need to meeting and/or exceeding a customer's needs and expectations. The definition of quality differs from one industry to another; however, everyone agrees that quality is a symbol of excellence and, for most, is a crucial key to success. Quality also has continued to evolve through the ages, as we moved from Stone Age skills to current, disruptive technologies and concepts.

### Quality Throughout the Ages: A Brief History

Throughout history, the pursuit of quality has been driven by a desire for excellence, efficiency, and customer satisfaction. While the methods and tools have evolved, the fundamental principles of quality management remain rooted in the goal of delivering value to stakeholders.[1]

**Stone Age (10,000 BCE–2,000 BCE).** During the Stone Age, humans developed basic tools and techniques for hunting, gathering, and crafting. Quality was likely assessed based on the durability and effectiveness of tools and weapons. Trial and error likely played a significant role in determining what materials and techniques produced the best results, with successful methods being passed down through generations.

**Ancient Civilizations (3,000 BCE–500 BCE).** Ancient civilizations such as Mesopotamia, Egypt, and the Indus Valley had rudimentary systems for quality control, particularly in construction, agriculture, and

1 Most of this section is based on materials from Banks, Jerry. *Principles of Quality Control.* John Wiley and Sons Inc.; 1st edition (January 11, 1989).

craftsmanship. In Mesopotamia, for example, clay tablets have been found documenting transactions and inventory, indicating some level of quality assurance in trade and production.

Egyptian, Greek, Roman, and various other civilizations produced many wonders. These wonders could be attributed to the existence of strict standards for mathematical precision, materials used, and workmanship. The Egyptians were responsible for constructing the magnificent pyramids meant to be the tombs of Pharaohs, or kings. Construction of structures like these involved attention to detail and emphasis on quality and excellence in architecture, measurement, and masonry. The Egyptians were also responsible for the development of the "365 days a year" concept and the further standardization of a year into months and days (Banks, 1989).

**Classical Antiquity (500 BCE–500 CE)**. Greek and Roman civilizations made significant advancements in various fields, including philosophy, science, and engineering. Greek philosophers such as Socrates and Plato emphasized the importance of excellence and virtue, laying the groundwork for ethical considerations in quality management. Roman engineers developed techniques for constructing roads, aqueducts, and buildings with a focus on durability and functionality, showcasing early principles of quality in construction. The Greeks enforced quality standards even in areas like art and literature, and the Romans were famous for building high-quality bridges, aqueducts, and so on. The Roman era also saw the increased use of reinforced concrete for a large variety of structures that were built. Strict laws for ensuring high quality and safety were enforced.

The Indus Valley civilization that flourished in the Indian subcontinent was known for high-quality city planning, spacious and well-planned buildings, and various other types of architecture. The fact that some of these structures remain standing today is proof of emphasis on the quality of material, measurement, and structural efficiency. Physicians of this era were adept at performing complex surgical procedures, all of which were standardized and recorded in books.

Quality control (QC) in ancient China reflected its society. The centralized autocratic state included a centralized system of quality control. The Zhou dynasty (11$^{th}$ to 8$^{th}$ century BCE) is said to have regulated handicraft industries to the point that each industry had separate departments in charge of storage of raw material, production, distribution of finished products, standards and quality, and so on. The main defect seen in ancient China's QC system was the fact that costs were totally neglected in the quest for high quality (Banks, 1989).

**Middle Ages (500 CE–1500 CE).** Quality control during the Middle Ages was often regulated by guilds, which set standards for craftsmanship and protected the interests of artisans. Guilds imposed strict regulations on apprenticeship, training, and product quality to maintain the reputation of their trade. The emergence of trade guilds in Europe contributed to the standardization of product quality and the development of apprenticeship systems.

The Germans had laws punishing traders who were responsible for poor quality. These extended even to traders selling consumables and food items. Germany was one of the first countries to advocate the use of trademarks and quality assurance seals. Products sold always bore a seal of guarantee from the producer. The Germans also pioneered the idea of corporations and had departments responsible for the quality of goods.

The evidence of the existence of quality planning control and improvement can be seen via the shipbuilding industry of ancient Scandinavia. The 4000-year period from the third millennium BCE to the first millennium CE saw the progressive building of more sophisticated ships owing to the high quality and standardization of design, tools, and materials used, as well as production processes.

The history of quality assurance in France dates back to the 12th century CE; cathedrals built by "guilds" were supposed to adhere to certain standards and quality of work. The weapons industry in France also is proof of adherence to quality. To control the quality of weapons, product control methods were developed and standardized. The suppliers of arms were carefully chosen based on their performance.

During the reign of Edward III of England, trading and manufacturing fraternities were reconstituted, ordinances were laid for the control of quality, and punishments were established for violators. England also pioneered the use of "hallmarking," which is a concept (in use even today) used in the determination of the quality of gold (in terms of carats). Products of gold and silver with the stamp of a leopard's head became the ultimate symbol of quality (Banks, 1989).

**Industrial Revolution (late 18th to early 19th century)**. The Industrial Revolution brought significant changes to production processes with the introduction of machinery and mass production techniques. Quality control became increasingly important as manufacturers sought to meet the demands of mass markets while ensuring product consistency and reliability. Pioneers such as Eli Whitney and Frederick Winslow Taylor implemented early quality management principles, including standardization of parts and processes.

Adam Smith, in his book, *The Wealth of Nations* (1776), introduced the concept of division of labor (or specialization). The Industrial Revolution also brought with it the idea of mass production, which meant that manufactured goods could be produced by dividing labor into small manageable bits. Due to the aforementioned concepts, the responsibility for quality was shared by many people. Benchmarking also proliferated during this era; competitors were striving to outdo each other by improving upon the other's designs and production methods. Organizations began to form during this period to make management more efficient.

**Twentieth Century**. The early twentieth century saw the emergence of quality management pioneers who developed statistical methods for quality improvement. The concept of Total Quality Management (TQM) gained popularity in the latter half of the century, emphasizing the involvement of all employees in continuous improvement efforts. Quality standards such as the ISO 9000 series were established to provide frameworks for organizations to implement quality management systems.

In the 1920s, the Western Electric Company sought more rigorous quality control methods that would engender confidence in their instruments and appliances. Walter A. Shewhart, an employee of Western Electric, introduced control charts to ensure proper quality control. Western Electric also introduced terms like consumer's risk, producer's risk, probability of acceptance, lot tolerance percent defective (LTPD), etc., that are associated with acceptance sampling. During this time, the use of inspections as a method of quality control increased, as did the advent of average outgoing quality limit (AOQL), operating characteristic (OC) curve, multiple sampling, and the demerit rating system.

The 1930s mainly saw the dissemination of Shewhart's ideas. International interest in quality control grew during this period. Joe Scanlon, in the U.S., put forth the concept of quality control through employee motivation and involvement. Elton Mayo conducted "Hawthorne Studies" that revealed that attention paid to workers stimulates quality and productivity, and a Joint Committee was set up for the development of statistical applications in development and manufacturing. It is worth mentioning that the British Standards (B.S.) were developed.

The system's approach to the formalization of quality started in the 1940s as the concept of teams evolved; teams of engineers and scientists were assigned to various tasks and were responsible for quality. After World War II, high demand for goods caused quality to suffer because of careless production, which led to the realization of the importance of quality, giving way to the creation of a professional association: ASQC (American Society for Quality Control) in 1946.

The American National Standards Institute (ANSI) and ASQC formally defined *quality* as "the totality of features of a product that bear on its ability to satisfy the given needs."

The 1960s were an era of total quality control, the concept of "zero defects," and the formation of quality circles. The U.S. Consumer Product Safety Act was also introduced.

W. Edwards Deming was instrumental in the dissemination and popularization of quality control in Japan in the 1950s and 1960s. Kaoru Ishikawa referred to the 1970s as a "companywide QC phase," which emphasized the involvement of every worker, from the company president to the machine operator. Later, Ishikawa and Genichi Taguchi carried forward the theories of Shewhart and Deming to make Japan a symbol for low-priced and high-quality goods.

The 1980s is generally regarded as an age of slogans and exhortations. Charismatic quality experts like Philip Crosby believe that slogans and exhortations go a long way toward spreading awareness about quality and improving quality. The focus during the 1980's started shifting from the concept of quality control to that of quality management.

Deming believed that, when it came to quality, a holistic approach by an entire organization to exceed customers' expectations was important. His teaching, the "14 Points for Management," became a core concept of Total Quality Management (TQM) and lean manufacturing principles that shaped the modern quality management industry.

**Twenty-first Century**. Quality management continues to evolve in the present with advancements in technology, globalization, and customer expectations. Concepts such as Six Sigma, lean management, and Agile methodologies have gained prominence, offering organizations tools and strategies for improving efficiency and quality. The digital revolution has enabled the collection and analysis of vast amounts of data, leading to the development of data-driven approaches to quality management.

Quality defined in the context of a product, service, or process meeting the needs and requirements of its intended purpose can be described as "fitness for use," a principle that Joseph Juran used to define quality. Juran is best known for the "Juran Trilogy," which consists of quality planning, quality control, and quality improvement. He emphasized the importance of management involvement in quality initiatives and the need for systematic approaches to quality management. Crosby is remembered for the "zero defects" methodology in quality management. He believed that quality problems stemmed from management practices rather than worker incompetence and emphasized the importance of prevention over detection of defects, advocating for a "do it right the first time" approach.

A visual depiction of the history of quality is shown in Figure 1.1.

## A Closer Look at the Quality Stages

The evolution from one stage to the next—the transformation of how we learn and practice quality—has been remarkable. History has witnessed many milestones, disciplines, and frameworks in each iteration of Quality through the years.

Quality 4.0, as understood in the industry today, represents the integration of modern digital technologies with traditional quality management practices. It marks a significant evolution in how organizations approach quality management, leveraging advancements in connectivity, data analytics, artificial intelligence, and other technologies to enhance product quality, operational efficiency, and customer satisfaction.

The evolution of quality management from basic ideas to advanced Quality 4.0 concepts reflects the progression toward increasingly sophisticated and technology-driven approaches to quality improvement. Each stage builds upon the foundations of its predecessors, incorporating new methodologies, tools, and philosophies to address the challenges and opportunities of its time. Quality 4.0 represents the latest frontier in this journey, promising unprecedented levels of quality, efficiency, and innovation in the fourth Industrial Revolution. Quality 4.0 is a journey of continuous improvement and adaptation to changing technological landscapes and business environments, drawing upon the foundational principles and practices established in previous iterations, as discussed below. With each phase, quality management has evolved, driven by advancements in methodologies and the innovation of new technologies. Abuzayed (2024) and Hamid, et al. (2019) provided comprehensive reviews of quality management by tracking its gurus and milestones over time, highlighting how principles, tools, philosophies, and techniques have evolved.

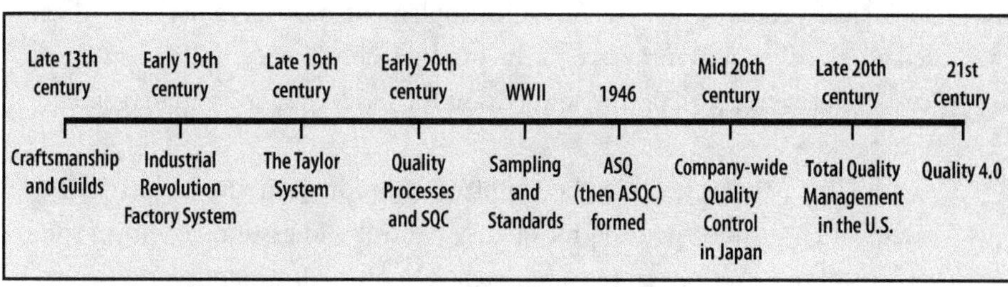


**Figure 1.1** History of modern quality.
Source: American Society for Quality, Milwaukee, Wisconsin, 2024.

### Quality 1.0: Inspection

Quality 1.0 emerged during the Industrial Revolution, focusing primarily on inspection and defect detection at the end of the production process. Inspection was typically performed by skilled inspectors who visually examined finished products to identify defects or deviations from quality standards. This reactive approach to quality management resulted in high levels of waste, rework, and customer dissatisfaction. Quality issues were often detected after production, leading to costly recalls and rejections.

The product itself received the focus through quality inspection and quality control methodologies. While this phase primarily focused on detecting defects through visual inspection and manual methods, it instilled the fundamental principle of ensuring product quality. Quality 1.0 was a reactive approach to quality management, and the need for earlier detection paved the way for Quality 2.0.

### Quality 2.0: Quality Control

Quality 2.0 was still focused on the outcome of the product but further analyzed the process of manufacturing for quality assurance. The concept of statistical process control (SPC) was pioneered by W. Edwards Deming and others in the mid-twentieth century, and Quality 2.0 emphasized the use of statistical methods to monitor and control production processes, aiming to prevent defects rather than just detect them. Techniques such as control charts, Pareto analysis, and root cause analysis were employed to identify sources of variation and implement corrective actions. That led to significant improvements in product consistency, reliability, and cost-effectiveness. It laid the foundation for modern quality management practices.

Quality 2.0 was born in the era of mass production and industrialization, which proved the need for quality control and assurance within the process, for a more proactive approach. Quality 2.0 expanded the scope of quality management beyond individual inspection to encompass the entire production process, fostering a culture of quality across organizations.

### Quality 3.0: Total Quality Management (TQM)

Quality 3.0 emerged in the late twentieth century, emphasizing a holistic approach to quality management that involved all employees and departments within an organization, with TQM as its main principle. Total Quality Management emphasized customer focus, continuous improvement, employee involvement, and process optimization as core principles, introducing many techniques and methodologies such as Kaizen, Six Sigma, and lean manufacturing to drive quality improvement and waste reduction throughout the value chain. TQM fostered a cultural shift towards quality consciousness and teamwork, encouraging collaboration and empowerment at all levels of the organization. This resulted in significant gains in product quality, customer satisfaction, and organizational performance. TQM became a cornerstone of quality management philosophy in many industries.

Along with TQM, Quality 3.0 witnessed the rise of globalization and the adoption of international quality standards such as ISO 9000. Quality 3.0 was widely focused on including the entire organization to ensure quality products were manufactured, further analyzing each part of the workflow process. Although still focused on the process and the product, Quality 3.0 paved the way for the entire organization to be involved in the quality system and showed how each person had a role to fill.

### Quality 4.0: Digital Transformation and Smart Manufacturing

Quality 4.0 represents the integration of digital technologies and advanced analytics into quality management processes, enabling real-time monitoring, predictive insights, and adaptive control. Quality 4.0 leverages technologies such as the Internet of Things (IoT), big data analytics, artificial intelligence (AI), robotics, and blockchain to optimize quality processes and enhance decision-making. It aims to enable proactive quality management, predictive maintenance, and continuous improvement through data-driven insights and automation. Quality 4.0 emphasizes agility, flexibility, and responsiveness to dynamic market demands, enabling organizations to innovate and adapt rapidly. Among its many benefits, Quality 4.0 offers the potential for improved product quality, operational efficiency, and customer satisfaction, while also driving cost savings and competitive advantage in the digital economy.

### The Emergence of Quality 4.0

Quality 4.0 emerged as a response to the increasing complexity and demands of modern manufacturing and business environments. The fourth industrial revolution, characterized by the integration of cyber-physical systems, IoT, and advanced data analytics, has paved the way for Quality 4.0. Key drivers include:

- The adoption of digital technologies across industries
- The availability of large amounts of data from various sources, including IoT devices, production systems, and customer feedback
- Advancements in AI, machine learning (ML), blockchain, robotics, and automation
- The need for global quality standards and reliable supply chains
- Increasing demands for high-quality, customized products and rapid service

Quality 4.0 is built on several core principles and objectives aimed at enhancing traditional quality management practices. Research in this area has been extensive. Ellis (2024), Jones (2024), Liu, Hu-Chen, et al. (2023), and Saihi, et al. (2023) have studied and identified articles related to Quality 4.0 concepts and definitions. Liu et al. (2023) have further "derived a general definition of Quality 4.0 and outlined four characteristics of Quality 4.0, which include quality management digitization, predictive quality management, mass personalization, and intelligent quality management." The main principles of Quality 4.0 include:

- Evaluating big data and advanced analytics to make informed and accurate quality decisions and improve product quality and process efficiency through actionable insights derived from data
- Employing IoT, AI, ML, and other digital tools to enhance quality management practices and increase automation, reduce human error, and achieve real-time monitoring and control
- Focusing on customer needs and expectations, and enhancing customer satisfaction and loyalty by delivering high-quality, customized products
- Shifting from reactive to proactive quality management through predictive analytics and preventive measures, preventing defects and quality issues before they occur, reducing costs, and improving consistency
- Promoting collaboration across the entire value chain, ensuring transparency in quality management practices, fostering a culture of continuous improvement, and building trust with stakeholders through transparent and accountable processes
- Implementing agile and flexible quality management systems capable of adapting to changing market demands and technological advancements, improving responsiveness to market changes and innovation, and maintaining a competitive edge

## Qualifying Quality 4.0

Many organizations aspire to implement Quality 4.0 to drive improved performance. Implementation efforts towards Quality 4.0 systems have largely resulted in incomplete systems that do not fully achieve the data-driven vision with which companies start. The major challenge facing most quality practitioners when given the goal of implementing Quality 4.0 is a lack of benchmarked success to follow. There is no set road map for Quality 4.0. Even defining what Quality 4.0 really means can be a challenge for many companies. The goal of this book is to provide clarity, explaining what the foundation of Quality 4.0 is and how to get started working towards it.

One of the most overlooked aspects of implementing a brand-new methodology is establishing a foundation for the concept. Quality 4.0 has become a popular buzzword in the quality profession, stemming from the term Industry 4.0. There are varied sources of information that may define these terms differently in subtle ways, and not having a clear foundational definition can create challenges when trying to develop a vision, strategy, and implementation plan. So, what is Quality 4.0? The most direct description is the marriage of quality and Industry 4.0.

Quality policies, management systems, goals, and costs vary among organizations. Deming, Crosby, and Juran all proposed different definitions for quality as well, so there is no standard definition that works for everyone. To define quality, an organization must be clear about its quality policy, goals, and vision. Before pursuing Quality 4.0, the quality goals must be clear, agreed upon, and well-documented.

The other piece of Quality 4.0 is Industry 4.0, which is the source of most of the confusion as a new concept. And since many publications refer to Industry 4.0 as a strategy or an end goal, it would help to know where the term came from. In the early 2010s, many organizations, such as the Global Economic Forum, recognized and documented the potential disruption caused by digital technologies (technologies that let us make use of data). We recognized that this collection of digital technologies would bring revolutionary changes to how we do business. Digital technologies were driving an industrial revolution: the fourth such era, to be specific. The first three eras were driven by steam power, electricity, and programmable logic controllers (PLCs). All these revolutions changed how we were able to do things, but they didn't change the value companies were trying to deliver—they just enabled that value to be delivered in new ways.

"Industry 4.0" was a buzzword created by the German government to describe their strategy of going digital. Industry 4.0 is a term that captures a group of digital technologies, nothing more. There is no correct way to go digital; there isn't one specific technology that must be used for Industry 4.0. Putting it all together, Quality 4.0 can be defined as the use of digital technologies to utilize data to achieve an organization's quality goals.

### Start with the Process

A key first step in rolling out new technology is to develop a good process for the technology's capabilities. Simply bringing in a software or other technology solution to solve an existing problem without first addressing the problem is likely to fail.

For example, navigation to a destination depends on paper maps, compasses, star charts, and experience. Today, navigation is a high-tech process with the use of global positioning satellites (GPS) easily accessible on any smartphone. The stated quality goals of the navigation process are likely:

1. Arrive at the correct destination.
2. Arrive at a reasonable time.
3. Prioritize safety along the route.
4. Utilize a cost-efficient route.

If all four goals are met, the navigation will meet quality standards. This process is a Quality 4.0 approach to improving navigation. The last ten years of advances in smartphone GPS have resulted in world maps that will actively update routes to avoid traffic, accidents, or tolls. The GPS used on phones actively collects data in real time from all the drivers using GPS to understand the flow of traffic and give meaningful information, presented in a usable way, to help a driver avoid an accident on the highway or traffic congestion.

The navigation process has been established and the supporting technology has improved, which is why it has worked. It makes sense that someone who can read a map would be able to adjust to turn-by-turn directions and become more efficient when those directions start using current data, but the opposite is not true. It would not work to give someone who has never driven on city streets or does not understand the nuances of driving the most advanced maps that have ever existed and expect that person to get to the destination. If that sounds like an oversimplification, consider taking someone from the rural Midwest who has never dealt with heavy rush-hour traffic and dropping them into Los Angeles with a GPS. Would they be able to successfully handle the "navigate in traffic" problem? Maybe, but they would not be nearly as successful as the California native who understands the process.

### Quality 4.0 Elements

The GPS example also shows the key pieces of a Quality 4.0 application. Quality 4.0 cannot be the same approach that was taken before the digital technologies of "4.0." The key difference between the paper road maps of trips past and the Quality 4.0 GPS is the data. In the example, data is collected from other drivers to support making decisions to drive higher-quality actions. The decision in the example is to change routes or do nothing; the data is all the positions and speeds of other GPS-using phones. But there is one more step in the middle: that data must be run through a model and presented in a meaningful way. It is not helpful to just see a point plotted with a speed vector for every driver on the road, and it is also not helpful if there is too much lag time between data collection and map update. The system must keep up with the decision making of the process. Therefore, a Quality 4.0 application possesses the following:

1. Data collected in real time or near real time

2. Data analyzed into actionable information fast enough to support decision making
3. Information presented to the right decision maker in the right format at the right time

These elements must be considered and planned for when developing a Quality 4.0 system.

## Five Key Steps for Implementing Quality 4.0

Every Quality 4.0 implementation will differ. Collecting data to enhance quality performance will depend on what data is important to the key value drivers of a company. How to collect that data will be unique, and how to analyze that data will depend on the needs of decision-makers. There is no single road map for implementation, but here are five steps that any organization can use, as shown in Figure 1.2.

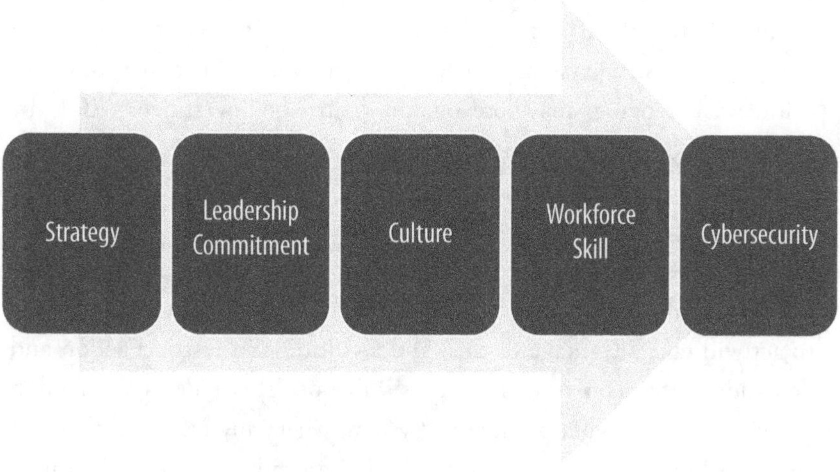


**Figure 1.2** Quality 4.0 implementation steps.
Source: Based on materials from Jones, Thomas J. 2024. "Qualifying Quality 4.0." *The Quality Management Forum*, 49: 4.

### 1. Strategy

Putting a Quality 4.0 strategy in place *before* starting a digital transformation is essential. Quality 4.0 may not be appropriate as a digital approach for all companies, so establishing quality as the focus of digital technologies should happen before any other planning. The other elements of a Quality 4.0 strategy are to decide which process will be the focus, what data will be collected, how will those data be collected in real time, what metrics or information will come out of those data through analysis, and who are the decision-makers who will receive the information. It may help to work backward from decision-maker to source data. With these strategy elements in place, a system can be scoped.

### 2. Leadership Commitment

Knowing the scope of the system is critical to understanding the cost and commitment required to implement the system. At this point, leadership commitment is required. Individuals will require training on new technologies, data infrastructure may need to be updated, time to implement and adjusting systems is necessary, and without patience and commitment, companies may lose out on return on investment (ROI) by giving up too early or training too little.

### 3. Culture

Quality 4.0 systems empower people with usable information to make decisions and act in real time. A culture that empowers team members to act will take advantage of this, and a culture that restricts action and demands permission from managers will not. This reflects the Andon principle of the Toyota production system, where any team member can stop the line when a problem arises. A lean culture of empowerment sets an organization up for Quality 4.0 success. Performing cultural assessments and working to adjust the culture if needed are actions that will make a difference in the future.

### 4. Workforce Skill

New technology requires new skills. The overwhelming number of technologies out there makes it nearly impossible to train the workforce on specific technologies. Once a strategy is in place and an organization understands which technologies it will use, it should commit to upskilling the workforce. This is key—all of the technology and systems in the world lose their value if the workforce cannot adjust to the system.

### 5. Cybersecurity

Cybersecurity is the single skills exception. Hiring cybersecurity skills at the beginning of a Quality 4.0 journey is worthwhile if a company anticipates going through a complete digital transformation. Cybersecurity fears are one of the main barriers that prevent top leadership from committing to Quality 4.0 today, and cybersecurity skills should be designed into these systems at the strategy level.

## Moving Forward with Quality 4.0

Once clear definitions and a straightforward starting point are established, Quality 4.0 success will be easier to achieve with everyone on the same page. Next, we will delve into the "why" of Quality 4.0.

# Chapter 2
## Why Quality 4.0?

INDUSTRY 4.0 WAS realized as the Fourth Industrial Revolution due to the rapid advancement in technology—and just as quickly, Industry 4.0 influenced quality management and quality control, as well as assurance policies within industries. With these technologies rooted in the principles of traditional quality management, Quality 4.0 was created. In other words, Quality 4.0 is the integration of digital technologies and advancements in quality management systems to enhance the overall quality of products and processes (see Figure 2.1). The technologies implemented through Quality 4.0 systems enable real-time monitoring, predictive maintenance, advanced analytics, and improved decision-making in quality control processes.

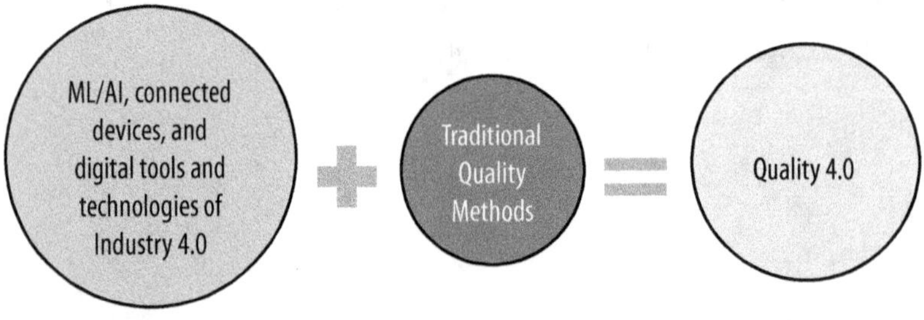


**Figure 2.1:** What is Quality 4.0?

Quality 4.0 embodies a transformative approach to quality management, integrating cutting-edge digital technologies to revolutionize processes across industries.

The main objective of this chapter is to outline the benefits and the impact of implementing Quality 4.0. First, we will present the components of a global Quality 4.0 framework, and then we'll discuss the technologies that drive Quality 4.0 implementation and the challenges that face such an endeavor.

## Components of Quality 4.0 Framework

Business analysis carried out by LNS Research (Jacob, 2018) identified three key components (people, process, and technology) with "axes" of a Quality 4.0 framework. Getting to grips with these axes, which include both familiar quality considerations and others that may be less familiar, will help your organization move toward the adoption of Quality 4.0 as the bedrock of your quality management systems.

As shown in Figure 2.2, and described on the following pages, there are three main components and their axes that comprise a Quality 4.0 framework (Jacob, 2017, and Juran Institute, 2019).

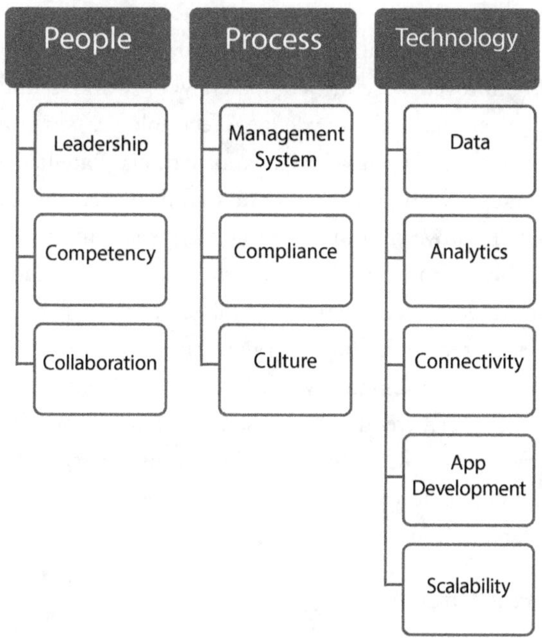


**Figure 2.2:** Axes of Quality 4.0.
Source: Based on materials from Jacob, Dan. 2017. *The Quality 4.0 Impact and Strategy Handbook*. LNS Research. Retrieved from https://generisgp.com/2018/02/15/the-quality-4-0-impact-and-strategy-handbook/.

1. People: leadership, competency, and collaboration
2. Process: management systems, compliance, and culture
3. Technology: data, analytics, connectivity, app development, and scalability

### 1. People Component

**Leadership**. Despite a widespread interest in quality improvement and the tendency to include quality as a corporate value, only 13% of companies say that quality is a priority for top management. To address this issue, quality teams must realign their objectives to link to the organization's strategic objectives, and quality leaders should advocate and lead quality across the organization, including at the executive level, to encourage buy-in and instill ownership.

**Competency**. Quality 4.0 can also facilitate a movement toward improved baseline competency of workers and better scaling of specialized knowledge. Technologies such as social media, machine learning (ML) and artificial intelligence (AI), mashup apps, wearables, and virtual reality (VR) should be leveraged to improve training and knowledge sharing.

**Collaboration**. Only 21% of the market has adopted an EQMS (Electronic Quality Management System), forcing quality to execute its (often collaborative) processes with the help of email, automated workflows, portals, and even paper documents. Tech trends such as social listening and blockchain have transformed collaboration in recent years, and leaders should look to leverage the transformative powers of connectivity, data, and analytics.

### 2. Process Component

**Management Systems**. Of those organizations with an EQMS, 41% have adopted a standalone, unintegrated solution. Due to fragmented core processes, companies often find it unrealistic to adopt quality technology. But harmonizing and automating processes and systems can allow quality staff to shift their focus from simply executing quality to innovating and improving it.

**Compliance**. About 36% of respondents in the life sciences reported that ensuring compliance was their top strategic objective for quality management, followed by reducing the total cost of quality. Quality 4.0 provides many opportunities to automate compliance. Today's tech providers deliver highly configurable, automated, and connected EQMS solutions, and even offer tools to automate validation. Companies should assess their current compliance systems and strategies to identify where improvements can be made.

**Culture**. At times, quality can seem to exist in a vacuum, despite its corporate-wide reach and importance. Only 13% of cross-functional teams clearly understand how quality contributes to strategic success. The improved connectivity, visibility, insights, and collaboration offered by Quality 4.0 make a culture of quality attainable.

### 3. Technology Component

**Data**. Data has always driven quality, and many updated regulations and standards highlight the importance of data-driven decision making. But most of the market lags in its adoption of real-time metrics, with only 22% of typical companies using real-time visibility of quality metrics in customer service, supplier performance, engineering, and manufacturing. Companies must seek ways to combine data from various systems to ensure accuracy and transparency.

**Analytics**. Thirty-seven percent of the market identifies poor metrics as the primary barrier to accomplishing quality objectives. Traditional quality metrics typically describe what happened, why it happened, and what might happen next, but they fail to prescribe what actions to take, an insight only available with the help of Big Data, ML, and AI. Companies are well-served to develop an analytics strategy after or concurrently with a data strategy; otherwise, the insights will be of little value.

**Connectivity.** Connecting business information technology (IT) (e.g., EQMS, Enterprise Resource Planning [ERP], Product Lifecycle Management [PLM], etc.) with operational technology (OT) (e.g., technology used in laboratories, manufacturing, and service) is a longstanding challenge. Of those that have invested in an EQMS, only 16% are connected to manufacturing operations. Industry 4.0 enables greater connectivity with inexpensive sensors that connect and provide feedback from people, products, and processes. Leaders should work to bridge the IT/OT divide, not only with technology, but by enabling data, processes, and people to work together.

**App Development.** Apps are becoming more versatile, customizable, and mobile. Recently, companies have created "mashup" apps that combine content from multiple sources into a single interface, a meaningful trend for quality with its involvement in all facets of operations and management. Leaders should consider the full potential of interactive apps available today, including wearables, augmented reality (AR), and virtual reality (VR).

**Scalability.** Thirty-seven percent of companies cite fragmented data sources and systems as a top challenge in achieving quality objectives. Without a global scale, traditional quality and Quality 4.0 alike are unable to reconcile processes, best practices, competencies, and lessons learned throughout the company. Cloud computing is of particular importance to scalability, along with data lake technologies. Start by assessing the current scalability—the ability to support data volume, users, devices, and analytics on a global scale—of your in-house systems.

LNS reported that participants in a survey by BCG (Boston Consulting Group), DGQ (German Association for Quality), and ASQ (American Society for Quality) acknowledge the importance of Quality 4.0 at all stages of the value chain. Their findings are listed here:

- Digital technologies can help improve quality in various ways.

- Monitor processes and collect data in real-time.
- Apply analytics to predict quality issues and maintenance needs.
- Enable people to do their jobs faster, better, and at reduced cost.

Other findings of the study reported that:

- Manufacturing and Research and Development (R&D) are the areas that will benefit most from applying Quality 4.0 by improving design and embedding quality into new products and capturing quality-related improvements in the value chain.
- More than 60% say that predictive analytics will significantly affect quality performance and the bottom line within five years, compared with only 16% who cite a significant impact today.
- By allowing companies to gather and aggregate data continuously, these technologies enable preventive maintenance and optimized production, thereby reducing the likelihood that a company will inadvertently release poor-quality products.
- Predictive quality, machine vision quality control, and digital standard operating procedures (SOPs) are the most important use cases in manufacturing.
- Companies can use Quality 4.0 technologies such as simulation testing and artificial intelligence (AI) to improve a design's robustness.
- Companies can work with usage pattern data that sensors connected to the Internet of Things (IoT) capture to improve the design of future products in ways that enhance quality by preventing failure.

- Digital tools for virtual design are important enablers of agile.
- Equipment in the field can communicate data regarding its condition to its manufacturer in real time via the IoT.
- Digital dashboards that track supplier performance give visibility to data such as key performance indicators (KPIs) and quantities ordered and received.
- By applying big data analytics to planning, a company can improve its ability to forecast the production volume it requires and to enhance the quality of its service.
- The most important factors involving cross-functional collaboration, according to survey participants, are centralization of quality data, an end-to-end quality management system, and quality cost transparency.
- Nearly two-thirds of survey participants believe that Quality 4.0 will significantly affect their operations within five years.

## Challenges of Quality 4.0

Like anything else, Quality 4.0 has its challenges. As part of the LNS Research study (Jacob, 2017), other studies (for example Kupper at al., 2019), present challenges organizations currently face in leading a Quality 4.0 digital transformation. Some of these challenges are listed below.

1. A shortage of skills is the main impediment to advancing in Quality 4.0. Soft skills, such as change management, communication, and teaming, are the most critical skills for success. There is also a need to improve analytics and big data skills.

2. An organization may lack a multifaceted approach that addresses the full range of strategic, cultural, and technological issues.

3. It is possible to create a virtuous cycle. If Quality 4.0 becomes a competitive advantage that drives business value, quality careers will become more attractive. Quality specialists will then benefit from additional resources, cross-functional responsibilities, and more competitive wages. Bringing in individuals with the requisite skills will take greater investment, and building those skills will take time.
4. Participants pointed to soft skills such as communication, teamwork, and change management as the most important skills for driving the overall success of Quality 4.0, stressing the importance of communication skills. One expert explained that Quality 4.0 requires "storytelling with data."

**What Does It Take to Lead the Transformation?**

You will need more than just technology to lead a successful transformation. Participants in the BCG study (Kupper, 2019) "regard soft skills (such as change management, communication, and teaming) as the most critical skills for success, even as they acknowledge the need to improve their analytics and big data skills." Findings from the same survey point to "the need for companies to accelerate their adoption of Quality 4.0." The study reported that successful implementation of Quality 4.0 requires a process of complete transformation of existing culture, technology, and strategy into a state of total digital transformation that includes mainly the components of Quality 4.0 addressed earlier.

**According to the same survey:**

- Thirty-seven percent of organizations identify poor metrics as the main roadblock to achieving quality objectives.

- Machine learning and AI insights allow prescriptive analytics to predict failure and inform what action to take to change the outcome.
- Twenty-one percent of organizations have adopted a core EQMS, leaving many businesses utilizing emails, automated workflows, and portals to manage quality activities.
- Thirty-seven percent of organizations cite fragmented data sources and systems as their main challenge in achieving quality objectives.
- Thirteen percent of cross-functional teams clearly understand how quality contributes to strategic success.
- Thirty-seven percent of organizations believe quality is critical for customer satisfaction.
- Twenty-six percent agree that the quality function has a clear and compelling role in delivering corporate strategy.
- Thirteen percent agree that quality is a priority for top management.

## Technologies Driving Quality 4.0

Quality 4.0 leverages several advanced technologies to revolutionize quality management practices. The key technologies driving Quality 4.0 and their impact are:

- Internet of Things (IoT)
- Data analytics
- Artificial intelligence (AI) and machine learning (ML)
- Blockchain for global supply chain
- Automation

Zonnenshain and Kenett (2020), Saihi, Awad and Ben (2023), Abuzayed (2024), and Ellis (2024) discussed how Industry 4.0 and

Quality 4.0 will introduce new concepts for quality as a data-driven discipline, incorporating innovation into quality management, and causing Quality 4.0 to emphasize data analytics and technology innovations.

**Internet of Things (IoT)**

Devices connected to the IoT, such as sensors and smart devices, continuously collect real-time data from various stages of the production process, providing to-the-minute insights into product quality. By monitoring the condition of machinery and equipment, IoT helps predict potential failures before they occur, reducing downtime and ensuring consistent quality, and therefore empowering the predictive maintenance capability of Quality 4.0. In addition, IoT sensors can monitor environmental conditions (e.g., temperature and humidity) to ensure they meet required standards, which is crucial for maintaining product quality in industries like pharmaceuticals and food processing. The IoT enhances information exchange, security, and administration. It intersects with product development, preventing repetitive innovation from competitors or internal sources. In addition, IoT simplifies data mapping in program development and enables group analysis. For example, it collects data on consumer spending habits and digital banking activity, such as identifying popular restaurants. It also helps designers determine product propensity in detail. Remote regulation of cooling and lighting products is possible through a network connection to the space.

**Data Analytics**

Big data analytics provide tools to help identify patterns and trends in quality data, enabling companies to address recurring issues and improve overall quality. Advanced analytics tools can quickly pinpoint the root causes of quality defects, allowing for faster corrective actions. In addition, analyzing customer feedback and complaints using big data analytics provides valuable insights into quality perception and areas for improvement.

### Artificial Intelligence (AI) and Machine Learning (ML)

AI and ML algorithms predict quality issues before they occur, based on historical data. This enables proactive quality management by employing AI-powered vision systems to automate the inspection process, providing more accurate and consistent quality checks than manual inspections. Exploring ML models will continuously learn from data to optimize production processes, reducing variability and improving product quality.

### Blockchain for Global Supply Chain

Blockchain technology ensures full traceability of products from raw materials to the end consumer, enhancing transparency and accountability. Blockchain can verify the authenticity of products, which is crucial in industries prone to counterfeiting, such as pharmaceuticals and luxury goods. It facilitates secure and transparent recording of supplier performance and quality data, improving supplier selection and management processes.

### Automation

Robotics in manufacturing processes ensures high precision and consistency, reducing human error and variability in product quality. Automation significantly increases production speed while maintaining quality standards, leading to higher efficiency and enabling flexible manufacturing systems to quickly adapt to different products and specifications, enhancing responsiveness to market demands.

The integration of these technologies creates a synergistic effect that drives comprehensive quality improvement. For example, the data collected by IoT devices are analyzed using big data analytics to provide actionable insights into quality management; AI algorithms process data from IoT devices to predict and prevent quality issues, optimize processes, and automate inspections; IoT devices ensure the accurate collection of data throughout the supply chain, which is securely recorded and verified using blockchain technology; and AI enhances the capabilities of robotics by enabling intelligent decision-making and adaptive learning in manufacturing processes. By harnessing these technologies, organizations can achieve a higher level of quality management, resulting in better products, greater efficiency, and enhanced customer satisfaction.

## Integrating Advanced Digital Technology

The integration of advanced digital technologies with traditional quality management practices is Quality 4.0. At its core lies predictive quality data analytics and ML algorithms, real-time monitoring through IoT sensors and other devices, supporting adaptive quality control, and implementation of automation technologies; together, these features promote data-driven decision-making, empowering organizations to analyze vast datasets, identify trends, and continuously refine quality management processes to meet and exceed customer expectations. The result is a range of benefits for the organization. Product quality is improved, operational efficiency is realized, costs are reduced, waste is minimized, customer satisfaction is increased, loyalty is within reach, and the organization is well-positioned against the competition. Each of the following benefits may occur via Quality 4.0 technologies [Ellis (2024), Antony, Sony, and Cudney (2020)]:

- **Improved Product Quality and Reliability.** Using predictive analytics, such as big data and machine learning, organizations can predict potential quality issues before they occur, allowing proactive measures to maintain high standards. Sensors and IoT devices enable real-time monitoring of production processes, ensuring consistent product quality by detecting and addressing deviations instantly. Blockchain and other digital ledger technologies improve traceability throughout the supply chain, ensuring that every component meets quality standards.
- **Enhanced Operational Efficiency.** Automation technologies, such as robotics and AI, streamline production processes, reduce human error, and increase throughput. Implementation of smart manufacturing systems

optimizes resource allocation and workflow management, leading to more efficient operations. Data-driven decision making with advanced analytics provides actionable insights, allowing for quicker and more informed decision-making processes.

- **Cost Reduction.** Predictive maintenance technologies anticipate equipment failures before they happen, minimizing downtime and costly repairs. In addition, IoT and AI help in optimizing the use of materials and energy, leading to significant cost savings and environmental benefits.
- **Waste Minimization.** Quality 4.0 tools support lean manufacturing principles by identifying inefficiencies and reducing waste throughout the production cycle.
- **Enhanced Customer Satisfaction and Loyalty.** Advanced manufacturing technologies allow for customization at scale, catering to individual customer preferences and enhancing satisfaction. In addition, digital tools, enhanced quality control, and faster response times lead to fewer defects and higher-quality products, resulting in greater customer trust and loyalty, and facilitating the collection and analysis of customer feedback, which allows for continuous improvement of products and services.
- **Competitive Advantage.** Implementing Quality 4.0 provides a competitive advantage to companies as industry leaders in innovation, achieves better market responsiveness and quick adaptation to market changes, and enhances brand reputation, making it easier to penetrate new markets and retain existing customers.

| **Benefits of Quality 4.0 Implementation** | |
|---|---|
| | Improved product quality and reliability |
| | Enhanced operational efficiency |
| | Cost reduction |
| | Waste minimization |
| | Enhanced customer satisfaction and loyalty |
| | Competitive advantage |

**Figure 2.3:** Summary of the benefits of Quality 4.0 implementation.
Source: Based on materials from Ellis (2024) and Antony, Sony, and Cudney (2020).

Figure 2.3 provides a summary of the benefits of Quality 4.0 implementation. Overall, Quality 4.0 adoption can significantly transform an organization's operations, leading to improved quality, efficiency, cost savings, customer satisfaction, and competitive positioning in the market. Radziwill (2018) highlighted that the fourth industrial revolution introduces seven tools and technologies to enhance quality in IT and organizations, focusing on how data should be utilized and governing information use. These are described here:

1. Data science and statistics help predict trends, develop models, and identify causal and non-causal links using data aggregation, categorization, real-time pipelines, and dynamic modeling for problem-solving knowledge.
2. Cutting-edge technologies like sensors, mobile devices, networks, IoT, Industrial Internet of Things (IIoT), integrated systems, virtual reality, and cloud computing are enabling connection innovations and are linked to document management.

3. Infrastructure for dealing with diverse, rapidly arriving data sets is essential (big data).
4. Constant monitoring and quality-based transactions are ensured by blockchain technology.
5. AI is used for complex judgments, such as computer vision, chatbots, and robotics.
6. Big data uses machine learning to aid in decision-making, prediction, information filtering, and recommendation systems.
7. Neural networks and deep learning predict and recognize intricate patterns.

## How Ready Are You?

Quality 4.0 incorporates the role of automation and digitization, and provides a competitive advantage for organizations by enhancing customer experience and increasing profitability. Antony et. al (2023) conducted a study to critically examine the organizational readiness factors for the successful implementation of Quality 4.0 and assess their importance. The study applies a quantitative research methodology to examine readiness factors of Quality 4.0 in organizations. 147 senior management professionals in various organizations, including manufacturing and service companies in America, Asia, and Europe participated through an online survey.

The study reported that "readiness factors for Quality 4.0 were critically ranked amongst manufacturing and service organizations by senior management professionals from three continents. Five significant reasons for the non-adoption of Quality 4.0 were:

1. The lack of resources such as technology, finance, and knowledge were voted as the first non-adoption factor for Quality 4.0 (28%).

2. Quality 4.0 is not linked to corporate strategy, and was voted as the second non-adoption factor (22%). This suggests that most organizations have not included Quality 4.0 as a corporate strategic initiative.
3. The third non-adoption factor is that benefits are not clear, and organizations find it challenging to evaluate the benefits of Quality 4.0 implementation (15%).
4. The fourth non-adoption factor has to do with the initial high investment for Quality 4.0 implementation (11%).
5. The fifth factor is that many organizations are already using other QM strategies and methods, and hence it is difficult to adopt Quality 4.0 (8%)."

The literature around organizational readiness and change [see for example Kupper (2019), Cervone (2016), Sony and Naik (2019), Santos et al. (2021), Hizam-Hanafiah et al. (2020), and Halpern (2021)] indicates:

1. Organizations must have certain readiness factors around leadership and top management support, strategic alignment of change, an organizational culture embracing of change, employee engagement and involvement, communication, knowledge about the fundamentals of Quality 4.0, training and awareness of and for the change initiative, as well as customer and supplier involvement.
2. Companies must make the digital transformation and Quality 4.0 strategy an integrated component of their corporate strategy. Also, leaders should wholeheartedly support these programs.
3. Organizations should help train employees in the new Quality 4.0 skills, and companies must install infrastructure that supports connectivity throughout the

entire supply chain while ensuring cybersecurity and the ability to assess their implementation of technology in terms of maturity and readiness.

4. Success in a transformational change, such as Quality 4.0, must consider the system as a whole and not in parts. It calls for a holistic understanding of the system and taking measures for its success (Antony et al., 2023).

## Benefits, Challenges, and Success Factors: Case Studies from the Field

**Benefits.** Several studies, field surveys, and qualitative interviews were completed to examine the benefits of Quality 4.0 implementation. They are reported below to examine the benefits, challenges, and critical success factors (CSFs) of Quality 4.0 implementation.

A qualitative approach was used by interviewing a global panel of senior management and operational excellence professionals from leading companies deploying Quality 4.0 (Antony et al., 2023). The panel reported that Quality 4.0 implementation within an organization improves customer satisfaction, ensures better and superior quality products and services, and delivers a competitive advantage. Radziwill (2018) reported that the new tools and techniques of Quality 4.0 enhance the quality of decisions by reinforcing human intelligence, ensuring transparency, and providing traceability. Further, it assists in robust prediction, unearths the bias, and dynamically provides feedback with potential solutions. Recent technologies such as blockchain have improved data quality and traceability, enabling efficient recall of products if quality issues arise. Javaid et al. (2021) reported that Quality 4.0 automates inspection processes in manufactured products, reduces the cost of quality, assists in dynamic quality control, increases the performance of the product, enhances business culture and partnership, re-aligns quality functions, reshapes production procedures, streamlines productions of new products, automatically detect defects, and helps in proper operations of finished goods.

**Challenges.** Quality 4.0 demands a systems thinking approach, quality information, reliable data, an innovative ecosystem, a learning organization, a lack of resources, a lack of alignment of Quality 4.0 with corporate strategic objectives, lack of awareness about benefits, and a high initial investment. The effectiveness of current continual improvement strategies compared to Quality 4.0 limits the adoption of Quality 4.0 across organizations [see for example, Antony et al. (2021), (2023), (2024) and Sony et al. (2020)]. Balouei et al. (2022) reported that the biggest challenge is to educate people—to enhance their technical abilities to solve the problem using scientific methodology. Several studies reported that organizations implementing Quality 4.0 need to understand the importance of the development of those technologies [see for example, Cudney and Keim (2017), Gunasekaran et al. (2019); Kannan and Garad (2020)]. Chiarini and Kumar (2021) emphasized the importance of leadership and digital literacy of people as required skills for Quality 4.0. Other studies highlighted a shortage of high levels of digital skills, technical skills, statistical and analytical skills, knowledge of data and decision-making, and problem-solving [Küpper et al. (2019), Sony et al. (2020)].

**Success Factors.** All studies mentioned above have agreed that for successful implementation of Quality 4.0, several CSFs should be readily available. To achieve the best implementation results, one needs:

- Top management support,
- A good organizational culture,
- Support of leadership,
- A vision for Quality 4.0 aligned with organizational strategic goals, and
- An effective training program in support of Quality 4.0 tools and implementation needs [Sony et al. (2021) and (2020)].

In conclusion, Quality 4.0 is a revolutionary quality transformation framework that, if properly implemented, could:

- Drive improvement across the value chain
- Reduce business costs
- Optimize business processes and organizational culture
- Increase the efficiency of operations
- Provide continuous improvement and innovation

# Chapter 3
## System Requirements for Quality 4.0

Quality 4.0 involves the marriage of Quality and Industry 4.0; Industry 4.0 utilizes data through modern technologies to drive decisions that are better informed to have better outcomes. This is not a new concept in the quality landscape, as data collection and analysis have long been a part of quality. Applications of statistical process control (SPC) and Six Sigma both provide tangible examples of historical use cases for data-driven quality control and quality improvements within processes. So, when thinking about how to implement Quality 4.0, it's not good enough to just think about using data, as that isn't a new concept; the Industry 4.0 side must be applied as well.

Historical use cases for data in quality are valuable for understanding how Quality 4.0 might look for a company. We can explore the foundations of SPC and Six Sigma to contextualize what quality data usage looks like. Many other data use cases demonstrate quality improvement, but exploring them all could be an entire text in and of itself.

### Statistical Process Control

Statistical process control, or SPC, has long been applied and taught to quality and operations practitioners in manufacturing. Montgomery (2020) described how SPC was originally conceived during World War II manufacturing efforts, as there was a need for more efficient quality control methods due to the increased manufacturing output needed to support the war effort. SPC is a tried-and-true approach to stabilizing production when the operating conditions of a process are established and known. SPC uses data to keep processes operating correctly by identifying instances when they are not operating in the way that is desired.

To leverage data to make decisions, SPC utilizes rational sample sizes and subgrouping to do a periodic sampling of a process parameter, to observe trends in a process, and to determine if that process is in control or drifting out of control. This method is more efficient than doing a 100-percent inspection on the throughput of a specific process, and it also can be applied to destructive test methods where 100-percent sampling would reduce throughput to zero. Statistical process control is a data use case, as the practitioner collects a statistically appropriate dataset, charts it into pre-established specification and control limits, and chooses to take the action of letting a process run as is (if data points indicate normal operating conditions), or stopping a process and investigating why a process has gone out of control.

With SPC, the key is to identify critical parameters of a process that must be kept in control. These would be parameters that, when they go out of control or change significantly, cause the process to fail. Consider the process of a morning commute. The process fails if the driver is late to their appointment (work, school, etc.). A critical process parameter is driving speed, as if the driving speed is too low, the driver will likely not be on time. Even if the driver leaves early to allow for slowdowns, as their speed approaches zero, the driver can be late.

With SPC, when a key parameter is identified, it is measured to identify statistical control limits. These limits are the statistical boundaries that would be expected when plotting a data point. There may also be specification limits, which are known limits of the parameter that would cause it to fail. With the commute example, if the parameter is "average driving speed," a specification limit can be identified as too slow, assuming the driver wants to leave at a reasonable time. A specification limit can be identified as too fast, assuming the driver wants to be safe and avoid being pulled over.

The parameter is measured over a sample size to calculate statistical control limits indicative of normal performance. To ensure success, SPC should be used to drive a process to operate within the range that will lead to successful outcomes. When control limits are calculated, performance can be monitored over time. The advantage of SPC is that when a parameter is graphed on a control chart, there are data rules that indicate when to investigate and potentially act. This makes controlling and improving performance much more efficient and effective, which is the goal of being data-driven. Suppose the average speed chart records a data point that is significantly over the upper control limit. In that case, an investigation may reveal that the driver started their commute late, which would lead to the action of setting a limit on the latest time the commute can begin for success.

## Six Sigma

Six Sigma is another popular data-driven approach to quality improvement in industry. Evans (2010) detailed how Six Sigma utilizes data differently than SPC in that a Six Sigma project involves creating a data collection plan and deploying statistical analysis to understand why a process is breaking down. Six Sigma can be used to define the key data points to monitor with SPC applications and is often deployed to solve complex problems within processes.

Six Sigma is applied largely through certification and mentorship by developing practitioners into project leaders who employ a data-driven framework to appropriate improvement efforts. Those efforts mostly target a reduction of variance. Six Sigma was created at Motorola in the 1980s and has been deployed to improve the consistency of processes with great financial success since it found its way into the industry at large. The Six Sigma approach to data is to collect process performance data and identify how capable a process is, with the name "Six Sigma" referring to a process with a capability index of 2.0, producing only 3.4 defects per million opportunities (DPMO). When processes are identified as lacking capability, those processes are investigated to identify sources of variance so that improvement projects can target them for reduction. The idea is that processes that don't produce defective parts result in more stable production and greater profitability for a company.

Six Sigma has been deployed by a long list of companies since its inception, including firms such as Texas Instruments, 3M, and Home Depot. Six Sigma also centers on the idea of identifying key process parameters that must be controlled. While SPC charts may come out of Six Sigma metrics, the Six Sigma approach involves taking on a phase-gated project management approach to identify and reduce the variance of key parameters. The phases of a Six Sigma project, historically, have followed the Define-Measure-Analyze-Improve-Control (DMAIC) process. Those are described below:

- **D**efine: Scope a project in the context of a problem and business case.
- **M**easure: Identify key process parameters and execute a data collection plan.
- **A**nalyze: Determine the statistical control of the parameters measured.
- **I**mprove: Target out-of-control parameters for actions to improve performance.
- **C**ontrol: Re-measure statistical control and make successful changes permanent.

Six Sigma is a powerful methodology when it is necessary to identify which process parameters demand actions and effort to improve the performance of a process overall, and it ties the improvement into business-level financial metrics as a second layer of data-driven decision making that drives value.

Statistical process control and Six Sigma are examples of different approaches to data-driven decision making that can lead to improved quality performance of a process. They highlight the value of using data and how Quality has a history of leveraging data.

## Quality Data Usage

As a general practice in both SPC and Six Sigma, data is fundamentally used to drive actions. In some instances, the action of "do nothing" is being supported through data, but it is important to note that this is still a data-driven action. Many businesses are focused on becoming data-driven, and SPC and Six Sigma are common methodologies used to achieve that goal. It is worthwhile to understand just how data is used in general. Consider the cases presented in this chapter; they begin with an end in mind, as the data to be collected is identified through analyzing the output performance of a process. The goal of data should be to identify weak points in processes, drive improved performance of said processes, and convert that improvement into business success.

Architecting a data use case should begin with value-adding actions in mind. In the case of Six Sigma, the use case of data is to identify sources of variance so that they may be targeted for improvement. With SPC, the goal is to identify out-of-control conditions to know when to respond to a process change. How to go about identifying value-added data use cases will be expanded upon in the next chapter on developing a data strategy, but to understand what Quality 4.0 systems require, it is important to frame the system in the context of value-added actions that are data-driven.

With Industry 4.0 technologies centering on data and being a critical element of Quality 4.0, it becomes increasingly valuable to understand how data usage improves quality performance in business today. New technologies do not require new practices altogether, and organizations should be leveraging best practices that already exist. Take logistics and shipping, for example. Humanity has always needed to deliver goods around the globe, which drove early maritime exploration. That need led to the Panama Canal construction. Exploration served an industry that now includes rail, trucks, and air freight in addition to shipping via boat. The advances in technology have empowered humans to be more effective at delivering goods, and we can now ship and package produce in such a way that, all over the world, it's conceivable to purchase tropical fruit at supermarkets. Technology didn't eliminate supply chains and logistics; it improved them. Data technologies should be used in the same ways.

## What Makes It 4.0?

So, the question to ask is: What makes a quality data use case 4.0? If it isn't good enough to use data in the ways that have already been established as best practices, what makes it better? This is where it becomes critical to extrapolate how Industry 4.0 technologies change how we can do things. It would be impossible to list all technologies under the Industry 4.0 umbrella, but they all improve our capabilities with data. At a high level, these technologies can be bucketed into the following categories related to data:

- Data collection
- Data management
- Data analysis
- Data usage

A snapshot of how these Industry 4.0 technologies are used in a Quality 4.0 system is highlighted in Table 3.1.

| Technology Type | Application | Example Technologies |
|---|---|---|
| Data Collection | Collects data by observing reality and creating a digital record | sensors, cameras, IoT |
| Data Management | Moves data from the point of collection to the point of analysis, houses and stores data for further analysis, accumulation, and usage | 5G, edge computing, databases, blockchain |
| Data Analysis | Converts data into usable information that can drive decisions and actions | AI, statistical software packages |
| Data Usage | Presents information to users to inform actions and decisions, driving the best outcomes | augmented reality (AR), user interface, connected personal devices |

**Table 3.1** Quality 4.0 technology elements.

## Data Collection

Data collection technologies are very straightforward; they empower the collection of data. This includes things like temperature or vibration sensors, which directly record a physical property that can be observed; cameras and audio-visual systems, which record images, sounds, and events; and other types of sensors, which may record more abstract data about the world around them. What these technologies have in common is that they are the starting point of data feeding a Quality 4.0 system, and they likely provide very little value as stand-alone technologies. Take, for example, a temperature sensor. This sensor serves the same purpose as a thermometer. It may seem like a temperature sensor can stand alone and be useful, but consider the intuitive example of wanting to know how hot it is outside to decide how to dress for the day. The sensor can be placed outside in the heat, and to get a reading, the user will have to go outside and experience the heat; that potentially nullifies the value of the system, as it isn't adding anything that the user wouldn't already gain by going out to feel the heat. The sensor can also be placed inside, where the user can see it, but then the sensor will be reading the (likely) comfortable indoor temperature so it won't be useful. To provide value, the temperature sensor needs to be paired with some other technology that allows the user to receive information without going outside. All data collection technologies have the same limitation. Cameras can collect incredible amounts of data, but without a software application or a human user to look at the visual information generated and make interpretations of it, the system provides no value. Data collection technologies are also broad in their applications, so it is worth researching the state-of-the-art in sensing technologies, as breakthroughs are being made frequently in this area.

### Data Management

Data management encompasses a broader range of technologies than data collection. Data management involves tasks such as moving and storing data. Advancements in digital memory efficiency and cost have been significant to the development of Industry 4.0 as a whole. Oftentimes data management includes local data storage, a data movement system, and unlocalized storage of data. Data will also continue to be managed after some degree of processing and analysis, and may also need to be managed long-term for traceability purposes.

For data storage, various types of drives and storage services exist. Solid-state storage is a technology advancement, as is cloud storage, which is enabled by advancements in moving data. Fifth Generation (5G) wireless connectivity has supported moving data quickly and effectively; other technologies, such as Bluetooth and Internet Protocol version 6 (IPv6), allow more devices to exist with internet connectivity, facilitating the movement of data through connectivity. Blockchain is a technology that serves the sole purpose of improving the traceability of information by creating a ledger that cannot be tampered with. In the background of data management is cybersecurity. Cybersecurity is a new field of study that is driving advancements in the safety of an

Industry 4.0 system. Managing data is a critical technological aspect of an Industry 4.0 system.

**Data Analysis**

Data analysis is the transition of data into usable information. The most popular technology to talk about in the Industry 4.0 sphere is AI. AI is all about data analysis, and AI itself is an umbrella term referring to several different analysis models. One of the most common AI approaches is pairing a vision system (camera) with an ML model (in which significant data is fed into an algorithm to identify patterns). Interpretation of images can be executed through advanced pattern recognition, so this approach is widely used and very successful.

Analysis may also be as simple as old-school statistics, and many new software packages exist to execute those types of analyses. In the example of the temperature sensor, implementing a decision tree that simply reads the temperature and then determines if it will feel "cold," for temperatures under 55 degrees Fahrenheit; "moderate," for temperatures from 56 degrees Fahrenheit to 80 degrees Fahrenheit; or "hot," for temperatures above 80 degrees Fahrenheit (depending on the preferences of the user of course); is a form of analysis. These technologies take data and convert it into a usable format. It may seem too simple or intuitive to use an example like temperature, but it is effective in illustrating the concept. Imagine a model that considers temperature, humidity, cloud cover, wind speed, and wind direction. In many places around the world, these factors all contribute to how an individual would like to dress on any given day. Seventy degrees Fahrenheit could be a hot day if it is coupled with high humidity, no cloud cover, and wind coming from the south, while 70 degrees could be very cold if the cloud cover is high and the wind is coming from the west (cold air coming down from the Rocky Mountains). Someone could always go outside and feel the day and decide how to dress, but a well-constructed analysis can provide that information to a user quickly and effectively to support decision-making.

### Data Usage

Data usage technologies are an interesting and exciting extension of analysis. There are two major types: the software user interface (which is focused on information delivery and intuitive guidance for the user of a Quality 4.0 system to realize as much value as possible) and automation. Three-dimensional (3D) printing is a technology that is very popular in the Industry 4.0 space, and it is a data-intensive technology. It involves automation to execute upon a (typically custom) design that may be technically complex to make with historical manufacturing methods. Another form of automation is a reactive vision system, such as the safety package on modern cars that will automatically brake or steer the vehicle if a potential accident is detected. On the user interface side, cell phone notifications serve as an effective example of how to present usable information to an individual. User interface design may incorporate standard work to guide users on how to utilize information that has been presented to them. Data usage is where the value in a Quality 4.0 system must be realized; the technologies applied at this phase must drive improved performance through action to make a Quality 4.0 system worth it.

## AUTOMATION IS A QUALITY 4.0 REQUIREMENT

Compiling all the elements of how data is managed and used to realize value clarifies what a true Quality 4.0 system is. After all, data collection is has been standard since modern quality control was first used in manufacturing. Manually collecting data by hand is not new, and sensors and vision systems even allow data to be collected passively without human intervention. Similarly, data has been moved and stored for many years as well. Libraries are just organized data and information storage facilities, if you think about it. Data has been moved using systems like the postal service, radios, and even carrier pigeons.

A Quality 4.0 system has data movement and storage built in, so again, data will move passively as it is generated. Historical analysis have been done on paper, with a user keying into an Excel spreadsheet, or in a meeting with the judgment of experts. In a 4.0 system, analysis is executed as soon as the necessary data is fed into the system. Data usage may still look similar in the user interface examples, as individuals receive information that has potential value and must act on that information. An effective user interface, as well as automation, will improve the speed and accuracy of the response to information that is generated. But the difference between modern uses of data and legacy cases is automation.

For a system to be a Quality 4.0 system, the data must be collected, managed, analyzed, and delivered for use under the umbrella of automation, and the tangible business value of that automation can be found in speed, consistency, and error proofing. Quality 4.0 systems don't do many things that are brand new; what they do is combine data usage best practices with modern technologies to generate usable information through analysis, faster, with fewer errors, and in more usable formats than ever before. Automation is a requirement for 4.0.

## Quality 4.0 Example

There are not many case study examples of Quality 4.0 systems currently in publication, as the field and the technologies are relatively new. To highlight what Quality 4.0 applications look like in practice, examples from predictive preventative maintenance and visual quality control can be used to illustrate Quality 4.0 realization of value.

### *Preventative Maintenance*

Preventive maintenance in a machine shop is one tangible example of Quality 4.0. A machine shop has the goal of producing on-spec parts, often with challenging or custom geometry, made out of high-value materials. A CNC (computer numerical control) or lathing application can be used for subtractive manufacturing, wherein a blank piece of material is inserted into the machine and different types of tools are used to trim away material. When working with challenging materials, the tools that are used to remove material must be very precise and highly engineered.

Take, for example, a very hard metal such as titanium; the tools used to cut titanium must also be highly engineered. These types of operations are known to wear the tools, at which point tools need to be changed or adjustments must be made in the cutting process to accommodate tool wear. If a part needs to be machined to within one thousandth of an inch of the designed geometry, then tool wear in the tens of thousandths of an inch can compromise quality. Once the material is cut away, it cannot be replaced, so wrong adjustments can scrap entire parts as well. In a historical machine shop environment, the machine will be set up for a production run; the parts will be run and quality tested at the appropriate interval; and periodically, the machine will be maintained with tools being replaced or adjusted. This process accommodates the wear on the tools and is designed to prevent sending out-of-spec parts to a customer.

Machine shop maintenance has been improved with Quality 4.0 by automating the measurement of produced parts with high precision. This data is then analyzed to see the trend of geometry changing on the parts as they come out. If a tool is worn, it is reasonable to expect cuts to become shallower over time. By analyzing the trend of how tools are wearing, the information on when to adjust or replace tools may be automated. The machine itself can make its positional adjustments to accommodate minor tool wear, and the system can actively communicate to the operator when it is time to change out tooling due to too much wear on the cutting tool. This creates business value by minimizing the risk of creating out-of-spec parts, as the greatest financial risk is the accidental shipment of off-spec parts to the customer. The system will optimize production by minimizing shutdowns for adjustment and tool wear to the absolute minimum necessary while performing 100-percent inspection to prevent sending out-of-spec parts (Xu and Hua, 2017).

### *Visual Inspection*

Visual inspection is another major opportunity with Quality 4.0. As recently as 2015, ML models have become at least as accurate as humans at identifying and classifying images. A camera paired with an ML algorithm can look at a part or product and identify visual errors as well as a human employee. The visual system also doesn't suffer fatigue; if an individual's job is to visually inspect parts coming from an assembly line all day, there is the risk of human error occurring as their eyes or mind tire during the task. The camera system can also operate faster than a person.

One example of Quality 4.0 vision systems in practice is a pasta plant. In pasta production, if the pasta is not dried appropriately, it may become flaky, look visually unappealing (a critical aspect of food sales), and possibly not cook properly. Pasta is made by producing dough from combinations of salt, flour, water, and other ingredients such as eggs; forming that dough into the appropriate pasta shape; and dehydrating it so that it may be cooked in the future.

In the factory setting, visual 100-percent inspection would be cumbersome and labor-intensive, due to the volume of pasta required to make a factory operate at a profitable scale. Instead, a camera collects visual data of the pasta, an ML model analyzes those visuals into usable information of "good" versus "bad" to categorize pasta, and then the system sends out information on which pasta to quarantine when necessary. This system realizes value simply by being feasible. The company likely wants to ensure that all pasta is visually appealing and will cook correctly, as a consumer brand and especially a food company must care about their customer feedback or risk their reputation degrading and sales falling. Human inspectors would struggle to handle the same volume of work that an automated vision system could, in addition to the higher risk of being inconsistent or making an error (Massaro, Manfredonia, Pellicani, and Birardi, 2019).

A tire plant employs a slightly different visual system. With tires, geometry is critical as they must all fit onto the vehicles for which they are manufactured—but rubber may also have visual defects that indicate weakness and a greater likelihood of failure. To make a tire, the rubber is formed into the correct shape, often in layers, and sometimes with other composite layers put in for strength. The side wall and the tread are both critical and undergo different manufacturing processes. Ultimately, all these layers and components must come together in a consistent and correct geometry, with sufficient strength and tread pattern to handle road conditions safely.

At the end of the tire line, the tires are measured for this geometry, and they are visually inspected to ensure that they don't have obvious weak points. A human doing these two tasks may be inconsistent, and taking geometric measurements of a donut-shaped product may be subjective—in addition to taking time, it may be difficult to establish how good or bad a product is. The Quality 4.0 vision system deployed for this application involves putting tires on a rotating fixture, spinning them, and having a camera observe the entirety of the tread and the side wall. Simultaneously, a laser is used to measure the geometry of the tire in space. The camera is paired with an ML model that determines if the tire is "good" or "bad," and the laser identifies geometry that is out of specification.

The system actively communicates to the quality professionals when specific tires require quarantine and for what reason. Value is realized from this system, again, due to feasibility. It is desirable to inspect 100% of all tires due to the potential risk associated with their failure. A catastrophic failure of a tire could compromise an entire vehicle during operation, which can easily result in loss of life. Minimizing this risk as much as possible realizes tangible value, as minimizing that risk protects the business from the potential reputational and financial loss associated with such a catastrophic failure. We would also say that most businesses and people whose work has such potential to cause harm if done wrong would be willing to invest in such a system for ethics alone (Massaro, Manfredonia, Galiano, and Contuzzi, 2019).

It's important to remember that Quality 4.0 architecture doesn't detail specific technology. A Quality 4.0 system doesn't require the use of AI, blockchain, or another particular technology. As long as the technologies deployed can collect data, manage data, analyze data, and apply use to that data, the system meets 4.0 requirements. Some common Industry 4.0 technologies are listed in Table 3.2.

| Technology | Description | Category |
|---|---|---|
| Blockchain | Digital ledger that creates a transactional record, used for information security, as the records cannot be edited or modified | Data Management |
| Artificial Intelligence (AI) | Computer algorithms that can use large-scale data to execute complex analyses | Data Analysis |
| Internet Protocol version 6 (IPv6) | Naming convention that increased the number of unique web addresses available, allowing for many more devices to be globally connected | Data Collection, Data Management |
| 5G WiFi | Wireless connection technology that allows much higher speed connections, empowering real-time applications | Data Management |
| Edge Computing | Hardware/software integration that pre-analyzes data near the point of collection to reduce the amount of data being sent on to for processing | Data Management, Data Analysis |
| Cloud Computing | Analysis and management of data at locations other than where it is collected or used | Data Management, Data Analysis |
| Software as a Service (SaaS) | Software provided by third-party organizations is used to process data through cloud computing within third-party systems | Data Management, Data Analysis |
| Simulation | Digital iteration of a system used to predict performance outcomes under given conditions to understand the impact of process changes | Data Analysis, Data Usage |
| Augmented Reality (AR) | Systems that present information to users interactively within the scope of their surroundings | Data Usage |

**Table 3.2** Example technologies by category.

| Technology | Description | Category |
|---|---|---|
| 3D Printing | Additive manufacturing that combines digital design, robotics capabilities, and interface materials to create complex objects | Data Usage |
| Industrial Internet of Things (IIoT) | Integration of data collection into shop floor devices for continuous monitoring of key process and product parameters | Data Collection |
| Big Data | High-volume data that can only be collected, managed, and stored due to advancements in technology | Data Collection, Data Management |
| Sensors | Devices that measure and record physical properties in objectively quantifiable ways | Data Collection |

**Table 3.2** Example technologies by category. (continued)

## Quality 4.0 System Deployment

Implementation of a Quality 4.0 system will be explored in depth in the following chapters of this book. From the foundational ideas of data usage and Industry 4.0 technologies, it is good to start with an idea of how to go about deployment. A proposed conceptual model of how to implement Quality 4.0 includes the following phases:

- Quality 4.0 strategy development (Chapter 4)
- Quality 4.0 technology architecture (Chapter 5)
- Quality 4.0 workforce and culture transformation (Chapters 6-8)
- Implementation of technology and realization of value (Chapters 8-10)

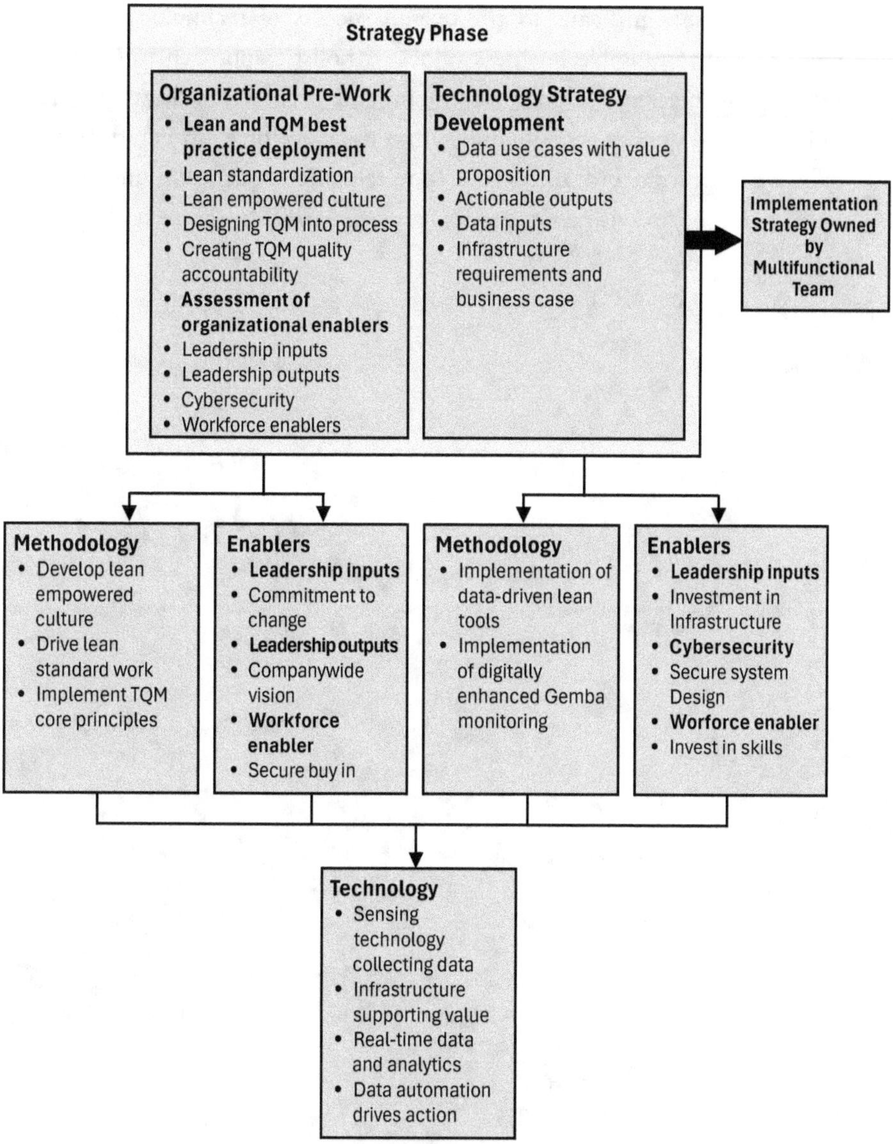


**Figure 3.1:** Conceptual model for implementation of Quality 4.0.
Source: Jones, 2023.

Figure 3.1 is an overall implementation model for Quality 4.0, with methods described throughout the book. The non-technical aspects of the model are significant to the overall success of technology deployment to realize tangible business value. A holistic approach to Quality 4.0 implementation includes developing an empowered lean culture, building total quality accountability into the workforce, and crafting an effective data strategy leading to a comprehensive technology architecture. You'll begin with a data strategy, as explained in the next chapter.

# Chapter 4
## Making a Data Strategy

A COMMON THEME IN the digital age is the idea that we should be making businesses smarter. Terms like "data-driven" and "decision-making" are used (maybe even overused) in the professional setting, and it's not always clear what that really means. Which data should be used? How should we use it? What decisions should be made with data? Are there decisions that can be made without data? It can get overwhelming to think about every decision that is made and all the data that might be collected—that's why a data strategy is necessary.

A data strategy is a critical element of success with a Quality 4.0 system. When "big data" first came into existence with the new technologies available, there were several new concepts around how we should be using it. Early ideas included the data warehouse and the data lake. A data warehouse is a method for storing mass data with a degree of structure and sorting, similar to storing goods in a warehouse. The data lake is the idea of flowing all data into a reservoir with no organization and just focusing on volume, much like damming a river to form a lake. These early ideas are still in use, but they both represent one of the paradigms we hold about data, which is that data projects start with data. This common-sense approach leads IT professionals to be at the center of any big data application, with a key focus on how the infrastructure and architecture of a data system can, and will, support moving and storing lots of data. While IT professionals and the work that they do are critical to success with any 4.0 implementation, bringing them in first isn't a data strategy.

One thing to remember is that successful businesses throughout history have been built without the mountains of data we can access today. The idea of collecting every piece of data possible and saving it all is a new concept, no doubt stemming from the age of business analytics and real-time data applications. We see the value of having data and using it, and that makes us think that flowing and storing data is the most important way to use these new technology applications. However, this way of thinking doesn't align with the business success stories that have existed long before the digital age. We'll present a handful of real-world examples in this chapter.

## Ford Motor Company

Ford Motor Company—specifically, its founder, Henry Ford—offers a terrific example of a company that made an industry (and the world) change its operations approach, generating incredible success. It is relatively common knowledge that Ford created the first assembly line to manufacture the Model-T automobile, but what may be less known is that Ford revolutionized wages as well. Ford actively chose to go outside the norm of paying the lowest price possible for labor when setting up the flagship manufacturing facility to produce the Model T. This was done for two reasons: One, he valued having the highest quality of employees and wanted to attract talent; and two, at the time, cars were a luxury only owned by the rich. Ford intended specifically to turn his workforce into potential customers by driving the cost of production down (the assembly line) and the available income up. This operational change revolutionized the American household and likely led to the norm as we know it today: cars that are accessible and owned by average American workers. The planning and detail that went into setting up an entirely new mode of production were incredibly in-depth, and the result was profound.

## Toyota

Another automotive manufacturer produced similarly revolutionary changes just a few decades later. Toyota, led by Taiichi Ohno, championed lean manufacturing through the 1950s and 1960s. This management approach is largely viewed as a toolbox of improvement methods in American industry today and is centered on a few core principles of single-piece flow-and-pull production. The driving idea was that making extra was wasteful, an important way of thinking in post-World War II Japan, where the economy was still recovering. All the very useful tools that we associate with lean manufacturing today were born of necessity to address specific problems that arose during the movement from batch processing to pull-based, single-piece flow production. In a pull system, items aren't made until they are ordered, reducing the financial risk a firm takes if it builds in batches and hopes to sell them later. Single-piece flow is the only logical way to achieve true pull, by making only a single item at a time as it's purchased. A culture of respect and empowerment is not specific to Toyota, but it is a mainstay in the Japanese work culture, and it is also seen as critical to the success of lean overall. A single-piece flow system must also maintain flawless quality, and this is much more achievable with an engaged workforce that actively takes ownership and pride for their work. Ohno drove lean throughout Toyota's production system and even upstream into the supply chain, which led to Toyota being the dominant automotive manufacturer in the world for decades.

## Walgreens

Walgreens is another company that found success before the age of big data. Walgreens chose to leverage the key selling point of convenience by understanding their customer base and choosing to expand beyond the competencies of being a pharmacy. They pursued locations that would offer the most convenience for their customers and began offering a drive-through option for pharmacy pickups. This was beyond the core competency of pharmacy, but it was well-aligned with what their customers wanted, and it drove decisions about store locations at the corporate level while they were growing through the 1980s and 1990s.

## The Value of Information

Countless other examples demonstrate companies making sound decisions without access to big data. So, what does the success of companies that didn't have intricate data systems have to do with data strategy?

None of those companies had access to real-time data systems guiding their every move and choice. What they had access to was information—and more importantly, information that informed their critical decisions. Ford knew the price point of a car and the average wage, as well as the potential for the assembly line to be more efficient. Ohno and Toyota knew that single-piece flow would reduce overall waste through inventory. Walgreens understood the needs and voice of their customer to identify opportunities to outperform their pharmacy competition in ways other than just pharmaceutical competency. Data strategies can, and should, be modeled after successful businesses in the context of how information is used. A common pitfall in today's analytics-driven atmosphere is the idea that data are valuable. Data are functionally useless. Yes, in this book about how to use modern data technologies, you read that right: data are functionally useless. What *is* valuable is *information*. The right information may be priceless. What makes a Quality 4.0 system valuable is the ability to turn data into information—but even more specifically than that, the ability to turn data into the right information.

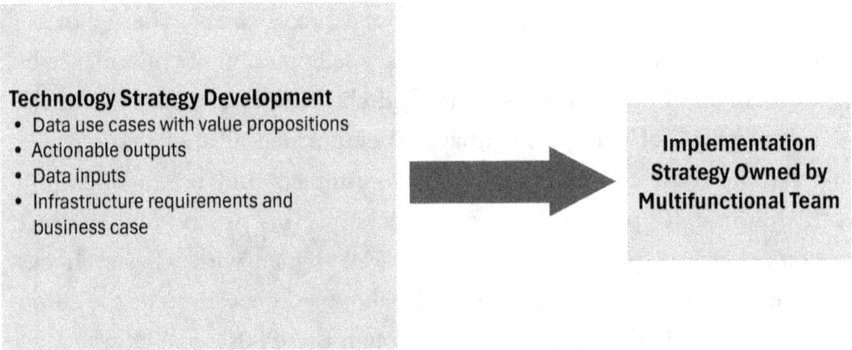


**Figure 4.1:** Technology strategy deployment section of conceptual model.
Source: Adapted from Jones (2023).

Identifying the right information is at the core of a successful data strategy. The right information empowers individuals to make the right decisions and take the right action, or, in some cases, choose not to take unnecessary action. Action is at the center of success, and the difference between taking the right action and wrong action can have tremendous implications on business outcomes. A sound data strategy centers on *action*. Data strategy starts with action because the outcome of data collection should be to act. Figure 4.1 begins with data use cases, which provide value. Identifying these use cases requires working backwards from value adding actions to the data that support them. Going from data to action is a five-step process:

1. Collect the right data.
2. Analyze that data to create the right information.
3. Empower the right decision-makers with that information.
4. Decide on the right informed action to take.
5. Act in the most informed way possible.

With action in mind, it becomes possible to work backward to create a data strategy. The process of creating the data strategy is to invert the process of executing the data strategy as follows:

- Select an action that impacts the value proposition.
- Scope with the decision that drives that action.
- Define the information that best feeds the decision.
- Design the analysis that creates the information.
- Identify the data that feed the analysis.

**Select an Action**

Selecting an action is the starting point of a data strategy. Not every action will be appropriate for Quality 4.0 applications, and many actions may benefit from 4.0 systems in novel ways that are easy to overlook due to the changing nature of how technology is used. The effort invested in selecting the right action will pay dividends, as actions that have high-value outcomes will help drive return on investment from digital infrastructure and accelerate the success of a business. Actions that are well-suited for Quality 4.0 solutions must contribute to the value proposition of a business. Digital systems may be cost-prohibitive or sufficiently complex to dissuade decision-makers from investing in them, and realizing tangible value from their outputs is the only way to justify them.

The value proposition of a business (or of a process, depending on the scope of the system) comes from the lean philosophy of value. Value is the output that makes a product or service desirable to customers, but it should be a description of what those customers need and want as an output. All actions taken by a business may indirectly support the value proposition in some way, but many actions are better connected to the value proposition directly. Those are the actions that should be the focus of a data strategy.

**Scope the Decision**

Scoping the decision may be more complex than the selection of the action. Many decisions get no real benefit from being over-informed by complex, data-rich analyses. It is very possible to choose a high-value action only to find that the decision to act isn't complicated at all; if that happens, the best path forward is to move on to another action instead of designing an over-engineered system for no real benefit. As a general rule, decisions driven by a Quality 4.0 system should be:

- Complex enough to benefit from data collection
- Resilient enough to withstand any decision being made too quickly
- Made at the point of action

## Decision Complexity

As already mentioned, not all decisions are complex enough to need a digital support system. The decisions that do benefit from digital support are not intuitive and may not align with typical logic. A decision that has very little risk if it is made incorrectly is also not likely to be a great Quality 4.0 application. For example, if a supervisor must decide whether to stop a line for maintenance when there is a customer deadline approaching, this decision is complex and risky; with no deadline, the decision no longer has significant risk. In the first scenario, it is important to be informed of all of the possible outcomes of continuing to operate or risking catastrophic maintenance failure. Identifying these decisions will be easier when including the decision-makers in the design process, as it's not always obvious which decisions in the value stream have the most risk if made wrong. Identifying the consequences of a decision helps to scope the complexity.

## Decision Speed

Any analysis that can be automated into a digital system can also be done by hand. The only difference between these two scenarios is how fast information is made available. Some decisions benefit heavily from decision speed, and others simply don't need to be made in real time. Decision speed is a major factor in focusing on the value proposition; many of the decisions that affect the value stream directly are the ones that have the biggest negative outcomes if they are made too slowly. An example of a decision that can be made slowly is the decision to begin development of a new product. Doing due diligence, collecting the voice of the customer, analyzing the business case, and many other steps can prevent a business from overinvesting too quickly in projects that may never make money. On the other hand, deciding to stop a production line benefits tremendously from fast access to information. The ability to respond to out-of-control operating conditions rapidly is the benefit of SPC, a method that has been used to improve business performance successfully for decades (see Chapter 1).

Another reason to focus on decisions that benefit from speed is the real-time nature of modern digital systems. Digital automation empowers us to move from the point of data collection, through analysis, to information, and to provide decision-makers with impactful information in seconds when the system is scoped correctly. Identifying decisions and their resulting actions that benefit from speed maximizes the real-time nature of Quality 4.0 systems.

## Decision Point

Realizing value becomes much easier when the decision point and the point of action are aligned. Taking real-time information and then creating a human-driven lag by providing that information to someone other than whoever needs to act on a decision reduces the value of the system significantly. Many decisions are best made with input from multiple individuals, and leadership oversight on decisions such as capital expenditure is deployed successfully in many businesses. These decisions benefit from analysis but don't happen fast enough to realize significant value from a digital support system. In the long term, these types of analyses can be built into existing digital infrastructure to realize extra value from the system—but at the data strategy phase, the focus should be on decisions happening directly at the point of value generation.

## Define the Information

Decisions are made all the time without picturesque reports and real-time information filled with insights and risk/reward analyses. These decisions, however, are very rarely made blindly with no context at all. Many times, these choices are driven by the experience of the decision-maker(s), which is just an undocumented historical record of the results of past decisions. This approach would also work with actual historical records, but those often miss nuanced details that come along with experience. Defining the information that supports a Quality 4.0 decision starts with identifying how the decision is currently made.

Documenting the information that goes into a decision in the current state is the first step. If a certain action/decision is always going well for a business, then there may not be significant ROI in developing a Quality 4.0 system to support the decision. It may be beneficial if the decision is specifically dependent on a small number of very experienced subject matter experts, and the goal is to empower many more people to make this decision successfully—but there is likely a risk of making the wrong decision. The second step in defining the information is to explore what information could exist that is currently available. This information represents the gap between successful decisions and failed outcomes. To identify the gap, it's useful to ask what extra information would have helped decision-makers in the past. Industry 4.0 technologies offer capabilities that haven't been available in the past, so it's critical to explore the unknown; developing a data system that only speeds up information that already exists misses a huge opportunity in the current technological landscape.

### Design the Analysis

The analysis is where we start thinking about data, and it's also where we must recognize that data, in a vacuum, are useless on their own. Analysis is the act of converting data into information, and there are several ways to do analysis. Something as simple as using a standard equation can be a form of analysis. For example, consider temperature. To Americans, 90 degrees Fahrenheit may feel hot, and they can decide how to dress accordingly when they know that temperature. If the temperature is measured in Celsius, they may not be able to gauge the weather properly. Knowing the temperature in Celsius is a data point that must be analyzed in this scenario, and there are two layers to the analysis. First, temperatures must be converted into a format that the decision-maker understands, and second, the context of experience informs the decision-maker what that temperature feels like to them. In this case, the information that drives a decision (choosing what to wear) is knowing if it will feel hot or cold outside. An analysis can be thoroughly complex, using ML or AI, or it can be relatively simple—but the important factor is that it must generate the right information.

The simplest way to architect an analysis is to ask the question of how the identified information can be produced. A key element of success in designing an analysis is the ability to quantify input factors as much as possible. It is challenging to build a truly subjective data point into an analysis. In the temperature example, the aspect of feeling hot or cold will vary from individual to individual. Certain objective criteria may be applied, but moving from subjective to objective data points is critical to analytical success. At this point in the data strategy, it may be worthwhile exploring the state-of-the-art in data collection, analytical methods, and benchmarking case studies, as technology may be able to collect data in ways that many professionals don't realize. Having a clearly defined analysis with objective inputs is the key output of getting to the foundation of a data strategy.

### Identify the Data

The starting point of using a Quality 4.0 system is the ending point of designing the data. At this point in the process, the analysis is well-defined, and an itemized list of data points exists. This is then the simplest step in creating the data strategy, which is to identify the source of each data point necessary for the analysis. The last step is to start exploring how the data will be collected, including identifying technologies that can measure the necessary data points. The technology side of collecting, connecting, and automating data will be explored further in the system architecture development phase of the Quality 4.0 implementation process. A data strategy is complete when all necessary data are identified to feed the analysis; that information drives the decision that results in the action that benefits the business.

The benefit of working backward through the process is having a clear connection to the value proposition. In many data system deployments, the project starts with data. Many professionals and subject matter experts can identify a lot of useful data based on their experience, but that doesn't create a clear link to value-creating actions. Collecting 80% of the right data may not be enough to support a thorough analysis and create the best possible information. Working backward ensures that the whole process stays on target. Another major pitfall of data collection strategies to avoid is focusing on what can be done with existing data and not exploring what new data need to be added. The best approach is to create a clear link to actions and business value, and to document a robust data strategy.

## THE FOUR VS OF DATA

When the desired data have been identified by relating data to actions, the second step in a data strategy is to ensure that the data are correct and accurate. Unfortunately, not all data are created equal. In many cases, a data point may be inaccurate from the source if it was collected inconsistently or if the measurement wasn't correctly designed. Data are the foundation of a Quality 4.0 system, so being successful with Quality 4.0 requires having good data as a baseline. In the quality world, it's common to say, "garbage in, garbage out." This phrase may have originated with raw materials in manufacturing; if the materials don't meet specifications, then the finished product will almost certainly not be acceptable. In modern manufacturing, this phrase can be applied equally to data. Starting with bad data gives bad information and drives bad decisions, and there are no exceptions to that rule.

So, how do we go about making sure the data is good? Start by looking at the four Vs of data:

1. Volume
2. Variety
3. Velocity
4. Validity

### Volume

Data *volume* is a straightforward concept. A data system needs enough data to not be biased. In statistics, sample size is incredibly important, and many quality practitioners deal with sample size math frequently in their day-to-day duties. Data systems are no different. There are two significant risks when dealing with data, along with one business factor, and they all relate to the sample size calculation. The first risk is saying something is good when it's actually bad. This is called the *consumer's risk* (or the type 1 error), which can happen when a small sample is biased or skewed. In the manufacturing world, this could occur when a factory receives one thousand parts from a supplier, tests the first one, and accepts the entire lot based on a good test. It's possible the lot could be 50% bad, in which case testing a single unit is a 50/50 risk of accepting a lot of bad product. If the factory tests two units when 50% are bad, the likelihood of testing both as good is .5×.5, or 25%. If they test three, the likelihood is 12.5%. It's easy to see in this simplistic example how testing more units is likely to give truer information to the individual who has to decide whether to accept or reject the shipment.

The second risk is called the *producer's risk* (or type 2 risk). This risk is the opposite of the first one, which is to say that the data say something is bad when it's good. Take the 1000-part example again; if only one part is bad and that is the only part that is tested by the customer, it may lead to a false assumption that the product is 100% bad when it is 99.9% good. It can be argued that a customer doesn't want to accept any bad product, but the reality is often that companies are willing to accept up to a certain percentage of defective products.

In both examples, testing all one thousand parts would give the most accurate information for a decision-maker to use, so of course that's what we should do, right? That is where the business case comes into play. Testing takes time and resources, and resources equal money. Some tests are also destructive, so if we test a part, it is no longer usable for production. Due to business considerations, it is rare to test everything all the time, so the type 1 and type 2 risks exist. Calculating good sample sizes is the best way to minimize these risks. This is the same concept that drives volume as an important factor for data. Too little and the system will likely give false results too often to be useful—too much, and the business case may not support the system financially. When it comes to real-time data, it becomes easier to have larger sample sizes; however, there is a cost consideration that comes with data volume. The cost of data analysis and storage scales with the amount of data being analyzed and stored. There are strategies to use bulk data and minimize storage costs, which will be discussed in the infrastructure chapter of this book.

## Variety

*Variety* refers to having data from a broad range of sources. This diversity in the data helps to create analyses that are more robust to data errors or misinterpretation. If data are being generated from a single source, then a failure of just one sensor or connection is enough to compromise the entire analysis. Multiple sources of data reinforce each other. A good variety of data also helps to maintain the system by surfacing errors. Some data points are not compatible with one another, and when they occur simultaneously, the data system can be set up to be able to detect and potentially even diagnose the failure.

Consider a data system that monitors environmental conditions in a controlled environment. The system could easily monitor the temperature with process control limits in place to detect process failures from being too hot or too cold. This could apply to chemical manufacturing, food production, plastics, pharmaceuticals, or any number of industry-specific applications. If the temperature sensor goes out of calibration and is reading incorrectly, the system may never detect the failure while the process is going out of control. Both type 1 and type 2 errors are possible in this situation. If the data collection system has a greater variety of data coming in, this risk could be reduced. In a pressurized environment, for example, temperature and pressure will increase and decrease together. A system measuring both is able to detect if sensors are failing based on the relationship. The most obvious example is when temperature is increasing while pressure is decreasing; in that situation, it's clear that something is incorrect. This can also be done by taking more temperature readings. Using three temperature sensors instead of one requires all of them to fail simultaneously and in the same way for the entire system to fail. Variety in the data being collected makes data collection systems significantly more robust.

Data variety must also be considered in the business case when deploying a system. Many data collection systems have significant differences in cost. It may seem like the path forward is to put every conceivable sensor on every piece of equipment and let AI run the operation—but there is a balance that has to be struck for a successful data system. The variety of the data being collected should support the data collection plan needed by the analysis in a cost-effective way that is feasible to implement.

### Velocity

The third V of data is *velocity*. Velocity and volume may go hand in hand, depending on the perspective you take. Velocity is all about having data that feed an analysis fast enough to be relevant and useful. If data move too slowly, then analyses in the digital age hold no advantage over data-driven practices of the past. The benefit of the technology we have today is that more data can be turned into better information through more powerful analyses, faster than ever before. Volume and velocity are central to realizing value from these technological advancements.

Velocity is the reason data collection requires automation to truly be a 4.0 system. The act of manually collecting data, putting them into a system, and then being able to proceed with the analysis already loses the ability for a system to operate in real-time. The real-time aspect of modern systems is the main selling point. Following the model of data becoming information through analysis, and that information drives decisions and actions, it is easy to see that the speed and quality of information is where the business value is. Using information to make decisions about customer desires is not new. Knowing what customers want informs the design process and future business decisions. The strongest endorsement from customers is where they spend their money, and companies have analyzed sales and market share for a long time; finding out what customers want after the product has been manufactured, distributed, and marketed is very expensive. Today, we have access to real-time feedback through reviews and social media. Companies can plug into customer wants right at the source by being active in collecting data online as soon as they become available. Being able to observe consumer trends as they are happening lets companies respond to changing market demands much earlier than has ever been possible. This example makes it easy to understand why velocity is key to the business case in data systems.

Implementing velocity in a data system is all about infrastructure. Where variety depends on the actual sensors and measurement systems chosen, velocity depends entirely on having the correct connectivity and data management. System architectures will be discussed in depth in the following chapter, but achieving the correct movement of data is one of the main goals of background system architecture.

### Validity

*Validity* is the final V of data. Data validity can be described as how truthful and accurate the data is. Validity is required for any data to be usable at all. If the foundational data going into an analysis is not true or accurate, it compromises the information generated as an output. It may be more damaging to make decisions with "garbage" information than to make decisions with no information or with only best-educated assumptions, as confidently trusting bad information may cause a decision-maker to remain committed to a fraudulent decision far longer than they would otherwise. They may also proceed down the wrong path with overconfidence in this situation. Validity of the data is how we combat this risk.

It should be self-evident that data validity is important. What may be less evident is how to establish it. Data validity is usually established by confirming the accuracy of measured data through parallel measurement systems that have already been validated. There may be a time threshold in which data points are confirmed through a third-party system to demonstrate that an automated measurement system produces data that are consistent and usable over time. Establishing the validity of the data coming into an analysis is a critical first step to any analysis, whether that be for a 4.0 system or not.

Establishing data validity should be one of the first tasks completed in the installation of a Quality 4.0 system. There must be a plan to establish that the flow from data to information produces positive results and generates value for the business, and establishing the validity of the data is the foundation of that flow.

## EXAMPLE OF A DATA STRATEGY

To bring it all together, let's walk through an example of what a data strategy should look like, following the actions defined earlier in this chapter.

1. Select an action that impacts the value proposition.
2. Scope the decision that drives the action.
3. Define the information that best feeds the decision.
4. Design the analysis that creates the information.
5. Identify the data that feed the analysis.

To follow through with making a data strategy, we must first identify a value proposition that we are supporting.

### Getting to Work on Time with Quality 4.0

A simple and intuitive example that can be applied to demonstrate the data strategy deployment is in the morning commute process example. This process is easy to describe and relatable to most people. The process that is being targeted for improvement is the morning commute. This process can be defined as follows: leaving the house, driving safely to the morning destination, and arriving. This process is successful, meaning the value proposition is defined as follows: if the commuter arrives safely and on time, it is successful; if the commute is unsafe or if the commuter is too late, it is not. For the sake of a well-controlled process, we can also define that the process is optimized if the commuter is not too early, as showing up well before the scheduled time results in waste, especially as time is a finite and precious resource.

The classic approach to solving this problem would simply be to try and consistently leave the house at the same time each morning, and that approach may provide the desired result for many commuters. To make this example more approachable as a Quality 4.0 application, let's consider some potential challenges to success. First, let's assume our commuter has factors which may make departure time inconsistent, such as a dog that needs to be walked and doesn't like rainy weather. Second, let's assume this commuter's normal drive includes crossing a railroad track and intersecting with school bus routes. Third, let's say that this commuter's normal traffic encounters have some level of variance, as occasionally drivers are moving slower than normal or faster than normal. Fourth, this commuter burns a tank of gas once every two weeks and may need to fill up on the way to work. Finally, let's assume that our commuter encounters construction about 5% of the time. Each of these factors can cause our commuter to be late.

So, plugging our now more complex process into the data strategy model might look like this:

1. Select an action that impacts the value proposition.

The most impactful action for this commuter is walking outside. In fact, in this situation, there aren't many other actions to consider at the surface level (more on that later). Going further down the process:

2. Scope the decision that drives the action.

The decision is what time the commuter needs to leave. If the action is scoped as it is, that is the only reasonable decision that will impact on the performance of the process.

3. Define the information that feeds the decision.

In this case, there may be multiple information streams that drive the decision. That may include knowing seasonal factors, such as school bus activity and train crossing times; random patterns of traffic variance; status of construction in the area; and fuel level of the gas tank. All these pieces of information can inform the decision about when to leave. Moving from the decision upstream:

4. Design the analysis that creates the information.

The analysis for this situation has multiple elements, and this is where moving from value backward to action, information, and analysis becomes very helpful. The easiest part of the analysis is the status of the gas tank. This is a single data point that may be directly converted into information. The data point is how full the gas tank is, and the information is whether there is enough gas to make it work or not in the morning. This analysis could be complicated by doing the math of the vehicle's average fuel economy multiplied by the amount of fuel and compared directly against the distance of the commute to call it a full-blown analysis, but the reality is that analysis only needs to be done once to determine a minimum level of fuel needed in the morning. Once the parameter of "minimum fuel" is established, the analysis becomes much simpler; fuel is either above or below the critical limit, giving a direct conversion from the data point of how much gas is in the vehicle to the information of whether gas is needed during the commute.

The next portion of the analysis is almost as simple but slightly more complicated: the construction determination. The construction data point is count-based and can be defined as "how many instances of construction exist on the commute today." Ninety-five percent of the time, the answer is zero, but it may also be one or more. The analysis in this case would require the data point of "How much time does the average instance of construction add to my commute," and that data point would be multiplied against the number of construction instances. This would indicate how much earlier the commute must begin to not fail due to construction.

The seasonal factors of train and bus routes would be analyzed in the same way as construction, but with one more layer of complexity. The simplest way to approach trains and school buses is to collect the data on when and where the buses stop, and when the train is using the intersection during the commute. These data points would have corresponding data points on how long the delay is for each instance. The third piece of the analysis examines what time the commuter would encounter these locations, based on what time they leave, and the fourth would be if the commuter and the bus stop, or the train, would be at the same place at the same time. This becomes a differential equation (and that is before variance is brought into play), but it is a feasible model to solve. The resulting information would tell us how much time would be added to the commute based on departure time on a given day; it would also be dynamic.

The final piece is the most random: the variance in speed and driving patterns of other drivers. To execute this analysis, we need data indicating the average number of drivers on the road in a certain time window and the variance in their average speeds. Opportunities to pass other drivers may also be considered, depending on the complexity of the commute. These data points would vary based on departure time, making them dependent on the other variables at play.

Looking at the situation through this lens, the idea of deciding when to leave for a morning commute is thoroughly complicated, and we aren't even done yet. That brings us to the last step in the Quality 4.0 data strategy:

- Identify the data that feeds the analysis.

Now, the specific numerical data needed to do the math has already been defined, and that will always be the case when architecting an analysis. In this approach, we focus on identifying how we will pursue that data. This is the technological approach. The data on traffic patterns and variance is impossible for the commuter to collect for individual cars on the road; however, there are options available (and this is an example of exploring the technologies available on the market) in the form of GPS systems offered by multiple tech companies. Simply plugging the commute into a phone will tell if the commute has traffic, as modern GPS offerings collect data from the drivers on the road that the commuter doesn't have direct access to see.

Similarly, many public transit schedules have apps and digital dynamic offerings that would tell the commuter when trains would be at the stops upstream and downstream where their commute crosses the railroad. This data can be collected through apps, but bus stops give us a potential challenge in the analysis model. At this time, we don't know of any systems that track school buses in real-time, so this data point may need to be replaced with an assumption of the likelihood of encountering a bus based on factors such as departure time and whether school is in session or not.

Construction data are likely also present in the modern GPS, but they may also be available online through websites managed by the jurisdiction responsible for the road being worked on. So far, the modern technology application of phone GPS and public transit apps provides a very convenient and automated method for bringing in the data needed, including automating much of the analysis. In the case of construction, the GPS will likely already dynamically adjust predictions for how long the delay will be, based on the combination of the construction with the traffic patterns. This is being done with ML models, meaning the Quality 4.0 technology application is already being offered to the average commuter through the phone service they already have.

Fuel is where the analysis gets more interesting. Modern technologies may automate part of the analysis, depending on how new the commuter's vehicle is, by calculating the miles available with the remaining fuel, and eliminating that step in the calculation. The real question, however, is how the commuter can collect the data on the status of the fuel tank before getting into the vehicle to start the commute. At this point, the commuter has already chosen a departure time, and if extra time was not planned for adding fuel, the commute has already failed. One way to collect the data is to depend on the modern technology of the cell phone and set a reminder for the morning when the evening return commute ends the day before. This would automate the data delivery on the status of the fuel tank at a time when the information can be used.

So, by using the GPS and cell phone reminders, the commuter can utilize Quality 4.0 to ensure 100% success with the morning commute, right? Nope. We forgot the finicky dog. Nothing about the dog's preference for not going to the bathroom in the rain can be controlled by the action of leaving and the decision of when to leave. The data strategy as defined is useful, but it isn't feasible to handle the dog situation. This gives good visibility into the holistic approach of viewing the overall value proposition and not focusing on data or details too soon. In the context of the value proposition, it may not be possible to scope a data strategy that accommodates all the potential failure modes of a process into a single action–decision–information–analysis–data workflow. In that case, it becomes important to revisit the data strategy. Fortunately, this second iteration is much simpler and provides an example of a quick and easy data strategy to implement.

The value proposition is achieved, in the context of the dog, by not departing late. Therefore, the value proposition for a successful dog walk can be defined as a precedent to a successful commute, and the dog walk value proposition is determined by it ending on time to not delay the start of the commute. The data strategy looks like this:

1. Select an action that impacts the value proposition.

The action is to start the dog walk early enough to finish it on time. Very straightforward. The assumption here is that the dog must complete its tasks for the walk to finish (this is a critical process parameter, as the entire day is a failure if we let our dogs down), and further, that the tasks take longer if rain makes the dog uncomfortable. This leads to the second step:

2. Scope the decision that drives the action.

The decision is what time to begin the dog walk—again, it's very straightforward. Next:

3. Define the information that best feeds the decision.

The information is simply the weather conditions outside. At this point:

4. Design the analysis that creates the information.

The analysis is the combination of whether it is raining, going to rain, or has previously rained. Finally:

5. Identify the data that feeds the analysis.

The information is a weather forecast, which can be produced by watching the news, checking a phone app, or even looking outside. For a Quality 4.0 defined process, the dog owner must consider looking outside to be out-of-scope. A good approach would be for an automated forecast reminder to come at a set time in the morning.

So, by determining how early to start the morning walk to ensure the dog can do her business without delaying the commute, and following through with leaving on time, based on the factors most likely to cause the commute to take longer than expected, the morning commuter can use a Quality 4.0 approach to ensure success with this daily process. The holistic approach to looking at the entire value proposition is worthwhile and should always be taken, as the ultimate goal of these systems, and the entire point of a data strategy, is to ensure that the actions taken lead to success in the value proposition. The bigger the process and broader the factors contributing to the overall delivery of value, the more complex and multifaceted the data system will ultimately become. This simple example can be broken down into multiple actions fed by multivariate analysis. An example of a full-blown production line may ultimately have dozens of individual actions that lead to success. The value proposition of an entire factory may have hundreds of actions happening all the time. An entire corporation may have thousands of actions driving success every day. What ensures Quality 4.0 systems directly contribute to this success is starting at value, working backward, and developing the data strategy in the right way.

# Chapter 5
# Establishing System Parameters

SYSTEM REQUIREMENTS ARE often the technical starting point for a Quality 4.0 system. As previously stated, this is where IT professionals and technical projects are executed—and usually, the technical requirements and technology business case are where these projects start. With the foundation of a strong data strategy to stand on, establishing these requirements becomes very straightforward and much easier to justify. Technology parameters directly follow the data strategy in their development and deployment (see Figure 5.1).

**Methodology**

- Implementation of data-driven lean tools
- Implementation of digital monitors

**Enablers**

- Leadership inputs
- Investment in infrastructure
- Cybersecurity
- Secure system design
- Workforce enablers
- Investment in skills

**Figure 5.1:** Technology methodology and enablers section of conceptual model.
Source: Adapted from Jones (2023).

Industry 4.0 isn't defined by one single technology; there are a multitude of possible options. Even AI, which is largely considered one of the most novel and revolutionary approaches to data use, isn't a requirement for any one 4.0 system. The best approach to scoping a system is to begin at the concept level before thinking about specific technologies.

The conceptual level system requirements can typically be broken down into three categories:

1. Data management
2. Data analysis
3. Information delivery

Data management describes the collection, housing, and security of data systems. Data analysis is the application of how that data are turned into the metric detailed in the data strategy. Information delivery encompasses the user interface and the aspects of designing systems to support the individuals who need the right information at the right time in the right place and format. Undergirding all these concepts is automation. Remember the Quality 4.0 system overview; there are many similarities to how we use data in 4.0 vs non-4.0 systems. Automation must be designed into a system to achieve Quality 4.0.

Detailing how data will be managed, analyzed, and delivered as information is the transformative process that takes a data strategy from use case to technical implementation. For example, saying that temperature will be monitored on a process that is sensitive to overheating makes sense and might be a part of a data strategy, but it doesn't detail the technical implementation of temperature monitoring. The hardware required to monitor temperature needs to be specified; in this case, maybe a temperature sensor of some kind that would be permanently attached at a critical point. The hardware also includes the connectivity and physical movement of data. For a permanently affixed sensor, it may make sense to hardwire it in; however, for a high-heat application, specialized cabling may be needed, as standard Category 5 wiring may melt. All of these details are important, and having a clear data strategy is critical to guide the detailed design of a Quality 4.0 system so that it is neither over-designed nor under-designed—so that it provides value while being cost-effective.

## Data Management

Data management specifically addresses the tasks of collecting, moving, and housing data. Cybersecurity should be included in the design of the hardware and software systems; when detailing system parameters, a cybersecurity professional should be on the project team. Some systems present more or less risk than others, but cybersecurity should never be overlooked. As cybersecurity is critical for all phases of system architecture, key cybersecurity concepts should be layered into all implementation plans.

## Data Collection

Data collection is not a new concept. With the design of effective metrics, the data that need to be collected are already itemized. The only real question on the system side is: How? For a Quality 4.0 system, most or all of the data collection should be automated. This requires knowledge of the state-of-the-art when it comes to data collection technologies. Recent advancements have increased the capability to automate data collection significantly. There are new sensors and new ways to physically install or connect devices, and advances in ML have created the ability for cameras to replace visual data collection and inspections. It is critical to design data collection plans with a knowledge of the technical capabilities on the market. Sensors may be connected wirelessly, whereas in the past this wasn't possible; they may be able to transmit data faster than ever before; and they may be capable of collecting data that couldn't be objectively measured in the past. This is all-important knowledge to design a Quality 4.0 system to deliver maximum value. Designing data collection is easy with the correct knowledge. The data points have already been scoped; finding the appropriate sensor or technology should follow a standard technology purchase/deployment approach.

## Data Movement

In addition to the sensors themselves, the hardware to move data should not be overlooked. The decision to hardwire or remotely connect different systems will determine the scope of data movement hardware requirements. Once those decisions are made for all data collection systems, data movement becomes very straightforward. The wired or wireless connection technology being used must be able to handle the volume and velocity of data being sent, and it must be capable of integrating with both the data collection and the data storage. In many cases, data movement may be integrated into other hardware. Bluetooth connectivity may be built into certain sensors or computers designed to receive data. Regardless of how data movement is scoped, it must support the velocity and volume requirements of the data system to be deployed successfully.

### *Data Housing*

Data housing is a more complex topic than data movement. Data housing requires storing data for analysis and use. This seems simple; however, the scope of the data and analysis may stress storage capacity. Vision-based systems, for example, may house video and images, which quickly consume large quantities of storage capacity. When computers are running at or close to the full capacity of their memory, they may begin to slow and become less efficient. Choosing how to house data is as critical to system success as the choice of which data to collect with which sensor packages. In some cases, a degree of pre-analysis may be necessary to reduce total data or to make a system more robust against noise. In the temperature example detailed earlier, the sensor itself may be capable of recording a data point several, or even hundreds, of times per second. Some of those individual data points may fluctuate with a lot of variance. If the temperature spikes up or down two, five, or even ten degrees for a second, is it necessary to capture all of those spikes? That depends on the specific analysis being done and why the temperature is important, but it is critical to ask those questions. It may make sense to send an average temperature reading every two seconds to the main system, significantly reducing the total number of data points. This may only make sense if the process itself has to be exposed to a sustained temperature change for issues to arise. Ultimately, the analysis as designed will require a certain scope of data input. Data housing must accommodate the needs of the analysis, and system storage parameters should be designed accordingly.

***Data Analysis***

Designing the data analysis is the central point of the data strategy. When scoping the system parameters, the analytical needs of a Quality 4.0 system will impact the scope of software, hardware, and integration of the system. Software selection must reflect the needed competencies of the analysis. This may include advanced statistical packages, in-house database development, software specialized to drive communication between different platforms, or any number of different competencies. The critical element is not to compromise the specific information generation required of the analysis through a selection of software and platforms.

Analysis software may be performed locally (often referred to as "on-prem" in the IT world, for "on premises") or digitally through the cloud. There is also an analytical method called edge computing. Determining which solution best suits the business case and value delivery of the system informs the hardware needs.

A cloud application will require data to be sent out for analysis. This may be a corporate cloud application, in which data are sent to a system that is off site but still housed by the company, or it may be a third party application in which a company utilizes software as a service (SaaS) and outsources analysis to a company who provides analytical and software services. With a cloud application, the data management hardware will need to be able to move a sufficient volume of data off-site with sufficient velocity to get results back in-house fast enough to be of value. Much of the development effort will be focused on moving data, security within an off-site system, and receiving results back from that data in a timely fashion. Advantages of cloud computing include not needing in-depth software competencies on site, while disadvantages include the risk of a security breach with data connected to an off-site system and the need to have robust software integration between multiple locations.

Local analysis is popular for companies that have software and analytical competencies or those companies that may not want to have their data connected to an online system for security purposes. In a local system, data is moved to an on-site platform and analyzed, and information is sent downstream as dictated by the value proposition—all on-site. This requires hardware to house data and software (such as a server room), upkeep of that hardware, and more physical integration locally. Performing local analysis may be faster and more efficient for those companies that have the competency in-house; however, it will cost more for data storage capacity than a cloud application.

Edge computing involves performing a partial analysis at the point of data collection. The term "edge computing" refers to doing some amount of analysis on the edge of the data collection system before sending partially processed data or information downstream for further analysis. These types of systems may be deployed to reduce the total volume of data to be moved from system to system, or to pre-process sensitive data into a lower risk format before exposing those data to the risk of a security breach between systems. The hardware and software requirements for edge computing are similar to those needed for local analysis, but they may require smaller storage and less capability, as information should move through an edge system; more intensive analysis will occur downstream. Edge systems can be paired with both more intensive local systems as well as cloud systems.

Hardware for data collection, storage, and movement, along with software for data analysis, and the integration between all those systems, is one of the more significant costs associated with the implementation of a Quality 4.0 system. This can often lead to the pitfall of cutting costs and therefore cutting competencies in the system architecture. It is critical to ensure that the system that gets implemented can support the analytical needs of the data strategy; otherwise, the system will fall short in realizing value and providing a return on investment.

## Information Delivery

While the challenges and costs related to implementation are heavily centered on the system hardware and software (the cost side of the equation), the value that can be returned by a Quality 4.0 system centers on the delivery of information. More specifically, as detailed in previous chapters, we need delivery of:

- The right information
- To the right individual
- At the right time
- To make the right decision
- To take the right action

The system side of information delivery needs to cover all these points to be effective. In quality, a common concept is to move quality efforts upstream. Identifying problem conditions early tends to reduce the total cost of poor quality caused by a problem, and that concept holds for a Quality 4.0 system as well. Collecting the right data may be costly, but analyzing them comes with a cost. Creating the right information has an associated cost, and achieving all those steps only to not effectively deliver the information is a very expensive failure. The entire system should consider the context of the user of the information and the use case of said information to avoid compromising the value delivery of a Quality 4.0 system.

### The Right Information

The right information should be well understood at this point in developing a Quality 4.0 system. It was identified during the data strategy, the analysis algorithm was developed to create it, and it provides the value delivery for a Quality 4.0 system. The right information is *what* a system needs to generate to be useful, and it must be in the form of information, not partially analyzed data that are only useful through the lens of a specific individual's experience or knowledge. Looking back at the temperature example: While an experienced individual may know that the machine being monitored reaching 100 degrees Celsius for a sustained period will cause problems, this is insufficient information as it fails to inform someone without experience what the condition is. In this case, the analysis should identify a time and temperature at which the information is clear and intuitive. It may be that the system indicates a machine is "at risk of overheating" under the right conditions, or that it has overheated. Even more effective is if the information includes recommended actions to take. It is important at this step that the "right information" is clear and direct so the end users can act accordingly.

### The Right Individual

In an ideal world, a Quality 4.0 system empowers people so anyone can become the "right individual." That may not always be the case; however, the benefit of a system like this is to empower people with information so performance can be improved with confidence. The right individual in a Quality 4.0 system is the person with the authority and skills required to act when information indicates that action is necessary. If the action to be taken requires management oversight, the right individual may be a supervisor or shift lead. If the action to take is in response to an immediate and critical failure, the right individual is most likely the person closest to the process, physically, who is most capable of performing an emergency stop or shutting the line down. The user of the information must be considered when scoping a system to maximize value.

### The Right Time

Getting information at the right time makes a tremendous difference in an outcome. The lag time between identifying a problem condition and adjusting or stopping a process results in running that process with the wrong outcome. This can result in rework or scrap costs, and, depending on the cost to run the process in question, those costs can add up quickly. In the context of automation, if a system is generating similar information to a traditional hand analysis, but the system is much faster, the financial benefit of that system can be quantified as the cost associated with the delay in getting the information from the hand analysis. Realizing this benefit requires the system to effectively deliver the information as quickly as possible to make a difference in the outcome. While it isn't a common problem, predictive analytics may run the risk of delivering information too early. Assuming the system is making accurate failure predictions (if the prediction itself is inaccurate, this would not be the "right information"), if that system were to deliver predictions too early, it may prompt the right actions sooner than needed, which could reduce the overall optimization of a system. This type of failure is not yet common; however, as predictive analytics are more widely adopted into use, it may be important to maintain awareness of when predictions are expected to come to pass.

### To Make the Right Decision

The "right decision" really is the culmination of the other "rights" on the list and ends in taking the right action. The key point leading to the right decision is that the information needs to be clear, direct, and relevant. The data strategy should center on the decision > action portion of improving quality performance, so at this point, the system simply needs to support effective information delivery. The system parameters are more relevant to the timing, individual, and having the right information, but the design of how to deliver those things needs to be informed by the decision itself and the action to take.

### The Right Action

The "right action" is the result of all the other "rights," and it is the value proposition of the system itself. At this point, achieving the right action should be an output of a well-designed system and a well-executed data strategy. Monitoring the outcomes of actions directed by a Quality 4.0 system is likely one of the best metrics to design to monitor the system itself, as both a validation of ROI as well as an opportunity to improve performance over time.

## The Right System Parameters

Delivering all the "rights" detailed in this section requires the user interface to be designed effectively. From a system parameter standpoint, the first decision is which system will deliver information. This may be a mobile device, it may be audible or an alarm of some sort, and it may be delivered via computer. How to deliver information will depend on the job scope of the recipient of said information. If the right person is often on the shop floor, away from a computer terminal, some sort of mobile device is likely the best option. In a facility that isn't too loud, information may be delivered over an intercom system of some sort as well. For a user who predominantly has access to a computer, email may suffice to deliver valuable information.

After scoping which system should be used to get information to the right person, the user interface must be developed. The user interface is what an individual sees and interacts with when it comes to software. The user interface must be intuitive enough to be effective at delivering the right information. This will require software and hardware integration to achieve. Once the delivery method and interface have been developed, the movement of information must also be integrated (either wireless or wired in). This is the final system integration to deliver information successfully.

## Cybersecurity Parameters

When dealing with any hardware, software, or connected system, it is critical to consider and design cybersecurity into the system. Cybersecurity concerns are one of the most common barriers preventing executive leadership from moving forward with Quality 4.0 systems. Cybersecurity is such a growing concern in the modern business environment that numerous certifications, master's degrees, and training offerings exist on the market. Cybersecurity needs to be designed into the system from the inception of a Quality 4.0 implementation project, and the best way to do this is to take advantage of those avenues to either hire or develop the needed skills. A cybersecurity professional is a critical member of an implementation team for Quality 4.0, and they should be hired or developed at the outset of such a project.

# Chapter 6
# Common Barriers

Alongside the goals and benefits described in Chapter 2, some challenges can prevent smooth digital transformations. This chapter details the common barriers impacting Quality 4.0 implementation and usage. It also addresses a path forward and presents recommended actions to be taken to resolve the main challenges.

## Acceptance Barriers

Many researchers found that the culture of an organization can be a barrier to success. This is considered a Quality 4.0 barrier as well, as work culture often serves as the foundation for all business decisions (Sony and Naik, 2020). Therefore, it is essential to foster an inclusive environment with motivation and empowerment. An example of how this can be accomplished is through a united organizational vision. Leadership can drive change by creating an agreed-upon mission, clearly communicating the rationale for change, and engaging early with stakeholders to act on the updated vision. This consistent collaboration will result in mutual understanding through providing personnel with a sense of responsibility while allowing leadership to maintain control as the traditional business landscape shifts to a more complex environment (Sony and Naik, 2020).

There are also apprehensions relating to human-machine integration and acceptance of these advanced technologies. Particularly involving the decision-making process, there is concern that human judgment is being replaced with algorithmically derived choices (Lewis et al., 2016). There are also concerns associated with both not trusting and overly trusting the Industry 4.0-enabled machines. Studies reiterated that this hesitation stems from the systems being viewed as unpredictable or having inexplicable functions. Related to this is another societal fear of the system developing its own agenda. Further research into this area of concern found that the fear stems from the innate human behavior and the survival mechanisms that can result in unethical actions. However, to address this fear, it is important to understand that Industry 4.0 tools allow for the opportunity for systems to be created without this predisposition; in turn, the system can potentially perform more ethically than humans (Novak, 2021). The key here is to understand the usage and benefits of the tools.

Another barrier to acceptance is the resistance to change. This could be a result of the learning curve required to adopt new processes or practices. It could also stem from the desire to continue using traditional systems based on the established operations and processes that have been functioning for years. To address these concerns, the organization needs to understand the technologies being implemented through proper training. Gaining knowledge on the benefits and appropriate usage of a toolset will help the workforce more easily accept the transformation initiatives. Further, involving the workforce early in the change process is essential for employees to directly observe the potential benefits while increasing knowledge of the Quality 4.0 solutions (Masood and Egger, 2019). For a more long-term solution, Sony and Naik (2020) recommend establishing customized training programs for current employees that focus on addressing knowledge gaps. In addition, a more formalized and strategic recruitment process can be implemented for new employees, which specifies the needs for explicit skillsets related to Quality 4.0 philosophies.

## Technology Barriers

Quality 4.0 focuses on revolutionizing systems and processes through embedding digital and physical systems. Because of this, technical components are essential to transformation. However, there are limitations with Industry 4.0-enabled technologies. As mentioned in the previous section, there are concerns involving the integration of the human and the machine, as well as challenges associated with trusting the systems. For example, from a usage standpoint, smart technologies have allowed for more realistic audio and video, which can potentially lead to unintended uses, such as forgeries, false reporting, or social discourse. Although there are initiatives in place to prevent Quality 4.0-produced falsifications, such as those from the Defense Advanced Research Projects Agency (DARPA), this vulnerability remains (Knox, 2020). A related concern is the system outsmarting forensic tools or the ability for the tools to be hacked. Gaining access for malicious purposes can result in intentional attacks on the system or data poisoning, especially with technologies that offer wireless capabilities. Possible solutions to these challenges involve incorporating all aspects of Quality 4.0, especially the approach of quality professionals transitioning from data analyst roles to data management roles, while continuing to monitor the systems for health and the data for high integrity and quality.

From an integration standpoint, there are also technical barriers that need to be addressed. If the new systems being introduced are not compatible with the existing systems, there will not be a successful transition. Further, if the new systems cannot fully replace the existing or required capability of current systems, there is an issue of not incorporating newer systems. To address this challenge, readiness assessments should be conducted before beginning implementation efforts. These assessments can include understanding needs and determining if replacements or modifications are needed for the identified technologies and processes. Readiness assessments are further discussed in Chapter 7.

Process compatibility is also another critical component that is often considered a challenge due to the need to determine which processes are value-added for Quality 4.0 solutions. To determine this, we recommend identifying and understanding the process before modifying and/or removing steps. The goal of the procedures should be to document or reference standards for the cyber-physical systems as well as the process of using them. It is also necessary to revisit the documentation for updates or improvements. Further, quality professionals should frequently assess the resulting performance data to ensure continuous improvement. For example, the real-time data can indicate Quality 4.0 principles should be added to streamline or enhance a specific product, process, or service.

## STRATEGIC BARRIERS

Implementing Quality 4.0 approaches involves the evolution of goals and capabilities from previous practices. This methodology enables quality practitioners to collect, analyze, and make informed decisions when given real-time data. An initial challenge during this change from traditional business practices is the need for a clear strategic approach to be defined by executive management. Managerial commitment can be shown through integration of the organizational goals with strategic planning, providing proper resources to stakeholders, and practicing the same methodologies that are being flowed. This approach should promote the incorporation of Quality 4.0 principles in all business practices and organizational disciplines. However, some studies found implementation to be challenging and costly across all levels and disciplines (Abollado et al., 2017). Others stated various recommendations for the implementation strategy that can be used to help with this challenge. Because the strategy is dependent on some factors, including the scale of implementation, it is important to assess the proper approach for your organization. For example, transformation can be incremental or radical (Pozzi et al., 2021). Incremental strategies are often considered an extension of the current business model, whereas radical strategies are deemed more innovative and revolutionary. The time and resources available to perform the transformation can help determine which of these approaches to take for your organization.

Transformation also involves an understanding of the required resources, the current organizational environment, and the enterprise-wide vision. The next chapter provides more details on how to create a strategic approach that addresses these constructs through the generation of an implementation plan.

***Resource Barriers***

Reduction of resources while incorporating Quality 4.0 is essential for sustainment purposes. However, a core barrier for organizational transformation is not understanding the required funding nor providing adequate time to perform associated tasks. Further interrelationships amongst these constructs and their respective challenges is reiterated with researchers agreeing that issues arise relating to human resources, such as the varying levels of skills amongst the workforce. Another example is the pivotal and influential role of managers and quality leaders. To establish and maintain this influence, proper resources must be allocated to directly influence and achieve the vision of the organization.

Investigating available resources before transformation is necessary to determine which additional resources are required to support the updated vision—or if specific resources are no longer needed. This can also help address the financial uncertainties and the ROI related to the new resources. To perform this investigation, we recommend conducting resource planning. Guidance on this assessment is provided in the next chapter.

## Data and Privacy Barriers

Related to the concerns of smart machines operating in unanticipated ways is the uncertainty involved with personnel and data protection. The real-time data and connections with the cyber-physical space allow for a new category of risks and increased vulnerabilities, which result in increased concerns about data security and privacy. Especially during the data collection and analysis process, there are challenges associated with potential bias, or the quality of data acquired from these data-driven models. To address these concerns, policies should be created and/or modified to aid in the protection of personnel, data, and the product (Sony and Naik, 2020).

There should also be an established cybersecurity policy in place before transforming the organization. Risk mitigation and controls should correspondingly be incorporated to aid in protection. Review data frequently for quality, timeliness, and availability before using the information for decision-making purposes. Because Quality 4.0 approaches are very data-driven, quality engineers should also monitor and review the data retrieval system for adequacy.

## Standards Governance Barriers

Many studies show the lack of uniformity in defining the constructs and goals of Quality 4.0 as well as the lack of standardized processes, procedures, and data governance. This causes challenges associated with acceptance, design, and usage of smart solutions. If they're not available in an industry, an organization should establish standards and policies for both designing and implementing Quality 4.0 solutions. The policies should require a common language across all components to allow for effective information exchanges. Across the business, there should be general recommendations for implementation, such as the creation of working groups or committees to help resolve issues as they arise.

Although most industries have specific ethical guidelines in place, the transformation of the entire business also involves evolving the ethical standards for the development and usage of smart technologies. This can help resolve concerns relating to the human acceptance of Industry 4.0-enabled systems and Quality 4.0 solutions. For example, the Department of Defense (DoD) requires the inclusion of personnel within the intelligent decision-making process to ensure the successful and ethical implementation of AI/ML (Lawless et al., 2020). Similarly, governance can include updating required communication efforts or recording the rationale for changes in processes.

These regulations need to be established and agreed upon before transformation efforts. The best approach to resolve these barriers is to standardize across all applicable organizational areas or industries, as this helps promote smoother implementation.

***Integrated Solutions***

As noted previously, many of these constructs are interrelated or impact one another. For example, through their updated vision, management can dictate the available resources through monitoring performance, metric tracking, and over-seeing funding; governance and documentation influence the workforce and their acceptance of Quality 4.0 through involving personnel in updating processes and early communication; and allocating resources involves factors such as technical readiness and the skills of the workforce.

Overall, to resolve the concerns associated with the main barriers of Quality 4.0 implementation, there needs to be ethical guidance in place for development and users. There also needs to be proper training or assurance of integrity through beneficial interactions between the human and the machine. Further, the standardization of policies, governance of data, assessment of required resources, and understanding of the knowledge gap are needed before the transformation. Even after the transformation efforts begin, quality engineers should use the available data to make beneficial changes for continuous improvement. These concepts are further expanded in later sections; however, the multifaceted implementation approach is very clear.

# Chapter 7
## Implementation Plan

QUALITY 4.0 IS imperative to business needs; however, resistance is often met when trying to incorporate new approaches. The common barriers discussed in Chapter 6 can be mitigated through developing an implementation plan based on the strategic vision of the organization and its business imperatives. A key element to Quality 4.0 success is the implementation strategy being owned by a multi-functional team.

This chapter offer six recommended steps for making an implementation plan. It's a whole-organization approach that acknowledges the need for consistent communication and collaboration across business levels and knowledge areas.

The six steps to establishing a strategic plan for implementing Quality 4.0 are as follows:

1. Define implementation success and intended outcomes.
2. Perform research and readiness assessments.
3. Understand the goals and barriers.
4. Conduct resource planning.
5. Perform risk assessments and risk planning.
6. Establish an implementation timeline and assign responsibilities.

Because implementation can be considered a series of strategically executed initiatives, the framework in this chapter follows the common project management sequence of initiation, planning, execution, and control (Sony and Naik, 2020).

## Define Implementation Success and Outcomes

It is important to define what success means for your organization so management teams and employees understand what is needed for success and what is expected of them. Defining success should be the first step in the implementation plan, as it sets the basis for the follow-on changes and activities.

To give some insight, the concept of successful implementation was researched, and previous studies reiterated that the definition of success depends on the application and the organization. For example, Masood and Egger (2019) claim success is the willingness of the organization to increase the use of Quality 4.0 and Industry 4.0 solutions. Other examples of indirect definitions of success involve establishing agreed-upon measures of performance. This can include the creation of additional products or reaching new target audiences. Another example definition of success is adhering to the planned schedule or delivery timeframe when using smart technologies in the updated process. On the other end, success can be a more direct and quantitative measure in terms of financial growth or sales.

In all these cases, the definition of success must align across all stakeholders and customers. The united perspective should then be used to create an updated vision and/or mission statement to be shared with the entire organization. These statements should be as specific as possible and should articulate the updated goals as well as the rationale behind them. Intended outcomes, or goals, are what the organization hopes to achieve or accomplish following the transformation. Havle and Ucler (2018) stated that the ideal outcome of implementing Industry 4.0 technologies is to increase competitiveness through enhanced sustainability and profitability. These desired outcomes helped to define organizational success as providing innovative, high-quality solutions at a lower cost. Masood and Egger (2019) used performance as an indicator of success, while Sony and Naik (2020) explained that the outcome of their success is greater levels of automation in their processes. (Additional outcomes and benefits of Quality 4.0 are explained in Chapter 2.)

Regardless of the definition, implementation success is accomplished through proper communication, adherence to the vision, and integration of all required stakeholders during the transformation process. Let's look at the recommended steps for achieving the successful implementation of Quality 4.0 philosophies.

## Research and Readiness Assessments

Part of developing an implementation plan is conducting research. In this case, it is important to understand the current state of the organization and its readiness to advance using Quality 4.0 philosophies and Industry 4.0 toolsets. Various literature provides examples of adoption models that focus on personnel, technology, and strategy; but the goal is to find the approach that best suits your organization, your transformational goals, and your definition of success.

To give some examples, Butt (2020) recommends using a modified Business Process Management (BPM) method, which first ensures all current business practices are effective. Multiple studies have shown that it is easier to transform a process that has previously adopted lean practices. This is an important consideration in determining readiness. Masood and Egger (2019) suggest following the Technology, Organization, Environment (TOE) method, where each measure can either promote or impede implementation success. Additional examples of approaches to investigate are the Technology Adoption Model (TAM) or the Unified Theory of Acceptance and Use of Technology (UTAUT). In general, a crucial component of these assessments is technology readiness, in terms of ensuring all systems are ready to transform or performing additional research to determine which systems are beneficial to transform.

When choosing a maturity model to assess readiness, it is important to find an approach that is not too narrow in scope and includes essential aspects of Quality 4.0, such as people and culture. It is also key to include aspects that can be defined as a barrier or a driver, as both can impact the implementation process. (Common barriers to Quality 4.0 are described in Chapter 6.)

Related to human-centric readiness is the ability of the organization or internal teams to be agile. The work environment that wants transformation should be flexible and adaptable to the changing needs of the business or customer. To deliver agile capabilities and speediness in results, Wang (2020) emphasized that a transformation plan involves more than just tools or infrastructure; it also encompasses changes in processes and people, where the latter is considered the hardest issue to tackle. This is also reiterated with the overall philosophies of Quality 4.0. It is more than embracing the future of quality and advancing technology—it is about transformation in every area of the organization. Because of these interrelationships, organizational readiness to transform is heavily dependent on multiple aspects, which should be addressed in the implementation plan.

The key aspects, which will be discussed in later chapters, include workforce involvement, resource allocations, and governance.

## Understand the Goals and Barriers

While developing this plan, it is important to understand the goal of establishing Quality 4.0 principles and the associated barriers. A common barrier is organizational acceptance of the new technologies due to the lack of understanding of potential benefits (among other barriers described in Chapter 6 earlier). To increase the chances for acceptance, early involvement from all organizations and levels is highly recommended.

As previously explained, providing a clear strategic vision and consistently reinforcing the updated mission is essential. Because leadership plays a pivotal role in innovation, personnel at higher organizational levels need to highlight the importance of transformation to their teams while being proactively involved with the changes. Management training should also be considered prior to disseminating information, as certain Quality 4.0

competencies and knowledge is crucial when leading the organization in advancement. Later in the process, a similar training program should also be provided to the workforce to help personnel understand the updated goals of the business and the defined measures of success.

Early involvement can involve proper training and communication among the workforce before implementation efforts. Neumann et al. (2021) stated that this involvement from the human dimension is essential for optimization, safe implementation, and employee acceptance. Further, because work culture often serves as the foundation for business decisions (Sony and Naik, 2020), it is important to foster an environment of motivation and empowerment. As an example, a case study involving multiple Italian manufacturing companies reiterated that success is found in organizations that aim to fill knowledge gaps through establishing proper training initiatives (Pozzi et al., 2020). These steps ensure employees understand the goals of transformation by understanding the use, purpose, and benefits of the technology being integrated. This also helps the workforce assess the impact on their everyday tasks and the potential benefits resulting from their human-machine interactions. These changes may otherwise be challenged if there is no workforce involvement in the integration process or if knowledge of the technologies being used, including their benefits, is not known.

To summarize this point, communicating the rationale for change and engaging key stakeholders are essential in the transformation and provide personnel with a sense of responsibility and control while increasing their willingness to accept the change. Further, communication and collaboration techniques result in a mutual understanding across all roles while the landscape of traditional business shifts to a more complex environment (Sony and Naik, 2020).

Lack of required resources (personnel or financial) is another challenge associated with Quality 4.0 implementation. Ensuring all necessary resources are available is vital for success. Therefore, the implementation plan should involve resource assessments and planning, as discussed next.

## Conduct Resource Planning

Quality 4.0 and Industry 4.0 focus on minimizing resources while maximizing desired outputs. To understand the impacts of this approach, resource planning is a key phase to be included in implementation planning efforts. Leaders need to assess impacts on current resources while researching what additional allocations are required to successfully implement Quality 4.0.

Research into various implementation efforts (described later in Chapter 9 with an example use case) identified four key aspects of resource allocation to consider: support from leadership, funding, time constraints, and required staffing. Although technology readiness is an essential component needed before beginning implementation, it is important to remember that implementation efforts are multifaceted and involve interrelationships amongst multiple aspects. For example, technical readiness also involves ensuring there is funding readily available to support the advancement of these systems. In addition, human-centric factors are often overlooked because of the overreliance on technological tools rather than the personnel in transformation efforts.

Overall, resource planning is another item specific to the organization, and there are a variety of software tool suites available to help with this phase. However, it is important to realize there is significant overlap and dependencies with the items outlined in this chapter. For example, investigations into resources should include evaluating the challenge of knowledge gaps and allocating a proper amount of financial support to help with this. General guidance is provided below with recommended steps to follow when creating a resource plan.

### Step 1: Identify and Inventory Current Resources.

This preliminary step helps the organization understand what resources are readily available. Examples here can include technical inventory, allocated financials, and human capital. Because a key component of Quality 4.0 is advancing by only using required resources, be aware of the current resources being used before transformation.

### Step 2: Determine Needed Resources.

The information from Step 1 can be used in Step 2. The list can provide a baseline for resources and be used to remove resources that are no longer needed or provide additional resources to fill any gaps. An essential component to making these decisions is understanding the defined goals and what is needed to reach them.

### Step 3: Ensure Integration of Resources.

Because Quality 4.0 implementation involves the transformation of an entire organization, the entire system needs to be compatible and able to integrate. For example, with technical resources such as computer systems, the various software suites should be configured properly and be able to communicate with positive interactions. In addition, the technologies should be flexible and facilitate communication through integrating physical products and the cyber world.

Integration of resources can also be human-centric. For example, integrating employee skillsets in the transformation process is key for sustainment. By understanding the knowledge areas of the workforce, tasks can be properly allocated, or a more effective training initiative can be established to help fill in the gaps.

### Step 4: Allocate Resources.

Building on the previous step, effective resource management aligns goals with the best-equipped resources to manage the defined needs. This may be aligning the intended outcome to a specific skillset that is required or understanding which needs require additional time and funding. Time is a key factor that is often overlooked when strategizing plans. Consequently, when allocating resources, the time needed to complete specific tasks should be considered. Interrelationships can be seen again as the time required for working assignments is also dependent on personnel and skillsets. Therefore, this step should also address interrelationships when determining allocations. For example, properly assigning responsibilities and overall resource planning help with employee retention and decrease the likelihood of burnout if expectations are established and strategic.

**Step 5: Track Resources.**

Creation of metrics can help track the progression of implementation efforts and may be used to document issues as they arise. Doing so can help resolve a concern before it becomes a costly problem. The earlier an issue is detected, the quicker a process improvement or solution can be incorporated, and make less of an impact on the bottom line. It is important to keep a pulse on task progression and performance, especially in terms of budget, to ensure there are proper resources allocated or highlight the need to remove identified bottlenecks. Further, the documented metrics, issues, and resolutions all can be used as guidance for future implementation and transformation efforts.

Using data obtained from the metrics can also shed light on areas of potential risks, such as unexpected tasks arising. To help with early resolution, systems and processes should be assessed regularly through improvement initiatives, including risk assessments, in case modifications are needed. It is also important to obtain and consider feedback from various stakeholders.

Use these steps for general guidance when resource planning. There are also numerous tool suites available to aid in this process. Some important features to consider when looking for resource planning tools include: the ability to obtain real-time updates, performance tracking and budgeting, smart decision-making features, and visuals of resource availability and team schedules.

## Perform Risk Assessments and Risk Planning

Risk management and contingency planning are often overlooked during implementation (Butt, 2020). However, these essential components should be performed and addressed in the implementation plan. Especially during transformation efforts, while the business environment is changing, it is important to minimize potential risks while developing mitigation strategies.

At the same time, we acknowledge that there will always be some level of risk in an organization or business, especially during transformational efforts. Not all risks can be managed or mitigated. All organizations must take on some level of calculated risks to be successful.

Risk matrices are a recommended tool to use for understanding the likelihood of a risk occurring and the associated potential impacts. This will also help with prioritization and development of strategies, since all risks are not equal; risks can vary, depending on their impact and severity. In most situations, the focus should be on the risks that can potentially lead to the greatest loss in value and have the potential for the biggest impact if they come to fruition. Developing a targeted strategy for managing high-level risks is essential for organizational success. Comparatively, emergent risks should also be considered. These can be identified through warning signs or trigger events. Risk matrices can help with all types of risk by tracking and assessing their patterns. The recommended steps to create a risk assessment matrix are as follows:

**Step 1: Identify Risk Categories.**

This step should involve collaboration among stakeholders to establish a foundation for risk assessment. Especially because risk analyses can be subjective, it is important to get agreement and input on the risk landscape from the beginning.

When developing categories, begin with the highest level of risks that often correspond to business functions, such as decisions at the strategic level. From there, the scope can narrow to specific processes, such as documentation management.

Examples of risk categories include strategic, operational, financial, and compliance-based risks.

**Step 2: Determine Risk Criteria.**

For most risk matrices, two criteria are considered: the likelihood of occurrence and the impact. Likelihood, or the probability the risk will occur, is usually placed on the x-axis of the matrix. Impact, or the level of severity of the risk if it were to occur, is usually placed on the y-axis of the matrix. As far as the creation of the matrix, the most common are 3x3 or 5x5, although this will be specific to the organization.

Like the first step, it is important to get agreement on the criteria that will be followed to evaluate the identified risks. This also helps facilitate future discussions related to mitigation.

**Step 3: Assess Risk Based on Criteria.**

Once the criteria from Step 2 have been defined, the risks should be assessed. However, to do this, a severity scale should be identified. There are multiple options for scales to use. For example, a simple "low," "medium," and "high" risk can be used in the matrix. Another option is the Likert scale of 1 through 5, with 1 indicating "extremely low risk" and 5 indicating "extremely high risk." Regardless of the scale used, the overall risk matrix methodology should be documented and used as guidance for consistency across organizations and teams.

Understanding the severities can then help the team determine the required resources and proper allocation to address the risks.

**Step 4: Prioritize Risks.**

The data obtained from the above steps, including the risk criteria and the risk scoring, can be used to prioritize risks. Prioritization should be based on risks that are identified as having the highest likelihood and impact. From there, you can create a risk assessment plan.

**Step 5: Create a Risk Plan.**

The plan can include control or mitigation efforts to either get rid of the hazard or minimize the risk to help lessen the chance of occurrence. This plan should also include a method to review or reassess the controls or mitigations that are selected to ensure effectiveness over time. Both the risk assessment and the risk plan should then be included in the overall implementation plan.

## Establish an Implementation Timeline and Assign Responsibilities

The implementation plan should provide a clear strategic approach to promoting and incorporating Quality 4.0 across all organizational disciplines and business practices. The approach should include the timeline for incorporation and all the events involved. To accomplish this, management must avoid over-scheduling or overloading their resources. Proper scheduling and planning should be completed in this phase of drafting an implementation plan. From an organizational perspective, and depending on the specific skillsets required, time to complete tasks may vary. From a process perspective, processes that have already gone through continuous improvement (CI) initiatives or adopted lean practices will have a smoother transition to Quality 4.0 approaches. This is important to keep in mind during the scheduling phase.

Similarly, the number of tasks and available resources are dependent on the available funding. In this step, it is important to outline a schedule by focusing on the most efficient use of resources, keeping in mind that early implementation efforts may take additional time. Scheduling based on availability also helps to more accurately predict the completion of the effort. With more precise predictions, the resources required for future demands are more accurately considered.

This step involves decision-making and compromising, if necessary, between competing milestones. In all these steps, especially this one, feedback is an important consideration. Requesting input helps avoid schedule conflicts or overbooking resources.

## Summary

Essential factors that should be addressed in the implementation plan include stakeholders, technology, and resources. To begin, leadership should clearly define the current state of the organization and where it aims to be based on the updated vision (Papke et al., 2020). This vision should include the specific definition and identified measures of success for the organization's transformation efforts.

It is recommended to use readiness assessments to assess if all components of the organization are ready to transform. A successful and smooth transition to Quality 4.0 can be challenging based on the associated barriers. Therefore, the goals and barriers should be identified and addressed within the implementation plan. Once these are defined, resource planning should be conducted and can be used to mitigate some of the defined barriers. It will also help with risk mitigation if resources are properly allocated. Risk assessments are essential during business transformations, as minimizing or mitigating potential risks can result in a smoother and more successful Quality 4.0 implementation.

Throughout all these phases, frequent assessment and reassessment should be performed to ensure the team is promoting the values of Quality 4.0. The use of metrics, resource assessments, risk management, and performance indicators can all be used to help with transformations and sustainment.

Like in the previous chapter, which discussed implementation barriers and emphasized their interrelationships, the multi-faceted topic of Quality 4.0 is evident in this phase as well. For example, resource allocations can be directly associated with risks and impact organizational readiness. This reiterates the Quality 4.0 philosophy on the need to transform the entirety of an organization to ensure the successful integration of advanced technologies.

# Chapter 8
## Realizing Value from Quality 4.0

THE TANGIBLE ASPECT of a Quality 4.0 implementation is in the system architecture itself. In terms of information technology (IT), the system architecture describes the hardware and software in place to collect, move, house, and use data in a 4.0 system. Many data system projects are initiated by IT groups, which results in projects starting with architecture as a main focus. This approach, while common, is backwards. It is the hardware/software equivalent of starting at the data instead of starting with the value.

One of the main drivers of a backwards approach to system architecture is cost. The hardware and software involved in a Quality 4.0 system represent the biggest line items when paying for Quality 4.0 implementation (along with training and development for the workforce). It makes sense to understand what data system architecture is already in place and how to maximize value from its use; however, if this is the only approach, the value of a 4.0 system may be compromised. Finding a balance between which existing systems can be leveraged against filling any gaps in the system is a critical step to achieve a Quality 4.0 system with a successful business case ROI. This process must start with the data strategy, as one of the biggest risks at play is starting with the existing system and compromising the data strategy to fit the system's constraints.

Once the data strategy is created, the existing system can be leveraged. In addition to seeing which aspects of a data system already exist, there is the possibility of modifying existing system parameters instead of building new ones. This is an opportunity for cost savings in the implementation phase of a Quality 4.0 deployment. Understanding the data collection and analysis needs and then mapping those to the existing system leads to the Quality 4.0 system architecture. An effective data strategy results in information improving the performance of a business in the context of the value proposition. To provide this value, the information generated by the system must outperform information that is already being generated. There are three ways to do this: the information must be better, the information must be faster, or there must be more information (or some combination of the three).

## Systematically Generating Better Information

Generating better information from the system comes from the analysis step. Not all analyses are created equal, and when developing a Quality 4.0 system, all of the potential of new technologies should be considered. New systems can use more data to get clearer results, they can process that data in more sophisticated ways, such as with AI, and automation will perform with more repeatability and less potential for human error over a static analysis done by a person. As described in previous chapters, a system that uses ML to interpret visual inputs can categorize information with greater accuracy than a human. This is a perfect example of identifying opportunities to generate better information. In this case, "better" would be defined as the accuracy of the categorization of an image. If a human inspector can look at a product and decide that it has no visual defects with a 94% accuracy rate, but the machine learning model can make the same decisions with a 96% accuracy rate, that 2% improvement of accuracy is the appropriate quantification of how much "better" the information is. When building towards an ROI, it is critical to adequately define a metric for the value of information; we will discuss that later in this chapter.

The second element of information being "better" is human error. When it comes to the analysis and management of data, there is a risk of typographical errors at every specific point where a human has to process, move, or handle data. This can be quantified by investigating the rates at which humans typically make errors and comparing that against the number of opportunities to make an error within the data handling processes that are used; it may be a good generic approach to quantifying the improvement of information through the reduction of errors. Risk can also be analyzed by monitoring current processes within a company and identifying the error rate that exists within the real processes of that company. With a repeatable analysis, a Quality 4.0 system will offer the benefit of reducing those identified error rates to zero, for the ROI calculation of the system.

### Systematically Generating Faster Information

In addition to producing better information, a Quality 4.0 system can create information faster. Much of the financial value and ROI will likely come from the speed of information. Better information being generated will require systems that go beyond the current capability of handling data. This is feasible; however, a skilled analyst can likely match the information quality of a system, even if they just have a static dataset in a spreadsheet, if they have enough time to thoroughly investigate the dataset they're given. If time is not a factor, an analyst will likely generate better information than a system due to the ability of the experienced human being to discern gaps in data sets, identify opportunities to add context, and do things as simple as reaching out to subject matter experts to ask for more information or context. This is an advantage humans will always have over systems, but it comes with the disadvantage of speed. The time required to investigate and seek out new information and add context will never produce information output faster than an automated Quality 4.0 system. Speed is where much of the ROI for a Quality 4.0 system will come from—especially during initial ROI calculations to realize budgetary approval.

Quantifying the value of speed requires an analysis of the output of the data strategy as it has been defined, specifically in the context of determining how much it costs to delay information getting to decision-makers. In the instance of defect detection, the value of speed is determined by factors such as the throughput of the system and the cost of containing a defect. In general, identifying a defect at the source and removing it from inventory, while simultaneously triggering intervention to prevent further defects, will be cheaper than catching a defect down the line. You won't have to identify and quarantine materials, and retroactively investigate a defect-causing process for potential intervention. In the case of the vision system, a camera may be installed within a production line in a place where humans can't safely work, which would empower the system to visually detect the defective part immediately instead of waiting for the part to advance further to a discreet inspection step within the process.

### Systematically Generating More Information

The Quality 4.0 system produces information that provides value another way: by producing more information. The amount of data that can be collected and analyzed is a limiting factor for quality departments and companies. This is not a new idea; the Pareto diagram is one of the seven core quality tools for this reason. With a Pareto diagram, the idea is to identify all defects that have occurred, rank them on a histogram, and allocate resources to solve the problems that occur the most often with the greatest impact. If resources weren't a limitation, this approach would be unnecessary, as companies would simply solve every problem and respond to every defect as they occurred. But due to resource limitations, only so many data can be collected, managed, or analyzed at any given time. A Quality 4.0 system allows for the amount of data collected and information generated to be increased significantly due to automation. This allows for potential reallocation of resources to better serve the organization as well.

Generating more information is valuable, as it reduces the need for stringent statistics and reduces the risk of sampling issues and bias on the overall interpretation of system performance. The sample size is the great equalizer in data; "more is better" is almost a universal rule of thumb when the goal is to use the data available to make the best possible decisions. The vision system example provides the most value by generating more information with its ability to execute a 100-percent visual inspection. This value comes from the context of the speed of the line and is related to the speed of information being generated. If the throughput is significantly faster than a human inspector, then a human-driven inspection will require rational subgroups and sampling plans. This always poses the risk of missing a defect or catching a systemic defect at some point after the initial occurrence. A 100-percent inspection built into the automated system eliminates both risks. This ROI is very tangible and can be calculated from historical performance and cost data; this is likely why early Quality 4.0 case studies include such a system and why they have already been deployed successfully.

### The Intangible Value of Information

The purpose of the information generated by a Quality 4.0 system ultimately centers on empowering the employees working within that system to perform at a higher level of quality. While it is not uncommon to read about how AI may be a threat to employees in the future, the Quality 4.0 systems described in this book will not replace the employees of a company.

The information improvements described in this chapter can be defined in the ways that overall performance will improve for the business as a whole—and the workforce should be taken into consideration, as they will be users of the system. Creating valuable information and putting it to use has the potential to drive change in the culture of a business, and, without changing the culture of the business, the financial returns from these improvements may not be realized.

The presence of faster, more accurate, and higher volume information should drive action. This, in turn, should lead to an empowered culture of people who trust that they may act when information dictates that action should be taken. One of the core tenets of a lean continuous improvement culture is that individuals are empowered to make incremental improvements. Systematically providing information to help guide these incremental improvements, while placing accountability for the accuracy of the information on the system and not the users, will create an atmosphere that organically grows the continuous improvement culture surrounding a high-value Quality 4.0 system.

Working within an information-driven system is also likely to drive more analytically-minded problem-solving into the workforce. Workforces are adaptable, and they are good at maximizing the system that they operate. A system that takes high-value data and returns high-value information to drive improvement in performance will create a culture of valuing information and data-driven approaches. As the economy becomes more and more data-centered, having a workforce that develops a high degree of competency in data systems and usage will help drive long-term success for a business. Developing a high-value Quality 4.0 system will impact the culture of a business to be more oriented towards continuous improvement, empowerment, and analytical thinking.

### Calculating the Financial Value of Information

The logic behind why Quality 4.0 systems yield better, faster, and more information has been discussed, but ROI calculations are tangible and specific. It is necessary to quantify the value of information with an actual dollar figure to establish the business case for investing in a new technology package. In all three cases (better, faster, and more), the value of information has to be established through the cost of not having that information. For this reason, not all information will ever be created equally.

To illustrate this point, consider the vision system that has been used as an example. If the system is inspecting the connection points on a printed circuit card that goes into a mission-critical system that leads to catastrophic failure if it breaks down—and that catastrophic failure could result in financial losses of over ten million dollars—missing just one defect can cost tens of millions. If that vision system is inspecting the connection points on a printed circuit card that goes into a child's toy that costs a few dollars and creates no safety issues with failure, a single defect can cost tens of dollars. The value of information is a million times greater in the first system, even if they are executing the same task. For the second system to be financially worth as much as the first system (assuming all elements of performance are identical and both systems provide the same accuracy and detection of defects), the factory making the children's toys would need to be producing millions of times as many units as the first factory. If the first factory is making aircraft, this might be the case. The value of information for the first system could be less, despite a single data point being worth a million times more, based on the context of the entire business. This example highlights the detail and context required to calculate the value of information.

Ultimately, there is no script for how to calculate the exact value of not having any specific piece of information that may be available anywhere in the world. Therefore, it is always useful to have professionals who understand capital projects, ROI, and the context of the business case involved in the planning and execution of a Quality 4.0 system. With that in mind, there are some consistencies in the value of information that serve as a good starting point. To calculate the value of better information, we must establish the cost of worse. For faster information, we must define the cost of slower. More information is contextualized through the cost of less information.

#### *The Cost of Current Information*

The cost of current information is likely the most challenging calculation in determining the value of the information being generated by the system. To make this calculation, conduct an analysis of the performance of current analytics and data usage to identify how much opportunity is left on the table. This involves looking at historical failure trends and putting them into the context of cost. This may also be a point within a Quality 4.0 project in which it becomes clear that the ROI is not yet cost-effective and the project should not continue—not because the system won't achieve the stated data strategy, but because the financial opportunity on the table with such a system isn't big enough to justify significant spending. This may be the case if the defect being targeted simply isn't costing the company much money, or possibly if the current system, which has worse, slower, or less information than the proposed new system, is performing at a high enough quality rate that there is no real financial gain to be made.

Calculating the cost of the current system involves taking the analysis of current performance and contextualizing it through the financial lens. It may be that:

1. The results of the current analysis could be improved to make better decisions, specifically if wrong decisions have been made based on current data usage.
2. The cost of lag between the occurrence of a defect, its eventual detection, and the quarantining and problem-solving that must occur due to the delay is high.
3. There is a lack of useful information, which creates a need for a follow-up investigation to make fully correct decisions.

***The Value of Better Information***

The value of correct information is in eliminating the cost of bad information. The easiest way to calculate this is by looking at the historical cost of poor information or errors in analysis and quantifying how many of those errors or costs would be addressed by the "better" information being generated by the Quality 4.0 system. Recall the vision system example: If the machine learning model is expected to be able to visually determine if parts are good or bad with 2% greater accuracy than the human inspector, this number should be weighed against the cost of the current performance. The current state is incurring two costs: the cost of type 1 (false-positive) and type 2 (false-negative) errors, which are the costs of false positives (saying that a part is good when it is bad) and false negatives (saying that a part is bad when it is good). The cost of a type 1 error is reflected in the rate of warranty claims and negative feedback from customers. This is the cost associated with the risk of packaging and shipping a bad product, and this cost is likely already known or analyzed by the quality department, as historically, quality has served the purpose of ensuring that a good product makes it to the end customer. The cost of a type 2 error can be described as the cost of the extra work to quarantine, analyze, and investigate a defect, as well as potentially the cost of scrapping or reworking a part in the instance that the false-negative incentivized the business to take those actions.

It is worth noting that simply multiplying dollar amounts by percentages may result in calculation errors. A holistic approach must be taken to ensure that the numbers reflect the reality of the business. To calculate the value of information in this example, the throughput, actual quality rate, cost of a type 1 error, and cost of a type 2 error must be known in addition to the improved performance of the information generated by the system. For the sake of an example calculation, we will make the following assumptions:

- The throughput of this production line is 50,000 units per year.
- A type 1 error costs, on average, $5,000, based on warranty data.
- A type 2 error costs, on average, $500 in extra labor and processing costs.
- The vision system is 96% accurate.
- The human inspector is 94% accurate.
- The first-pass yield of the line is 98%.

With those assumptions, we can calculate the value of information in improving from 94% to 96% defect detection accuracy. We will also assume that both the human inspector and the system classify both good and bad parts with the same accuracy level. The actual production performance of this system is outlined in Tables 8.1–8.4.

| Total Parts | Good Parts | Bad Parts |
|---|---|---|
| 50,000 | 49,000 | 1,000 |

**Table 8.1** System performance.

The production system itself operates at a 98% first pass yield, meaning that of 50,000 parts produced annually, 49,000 will be good and 1,000 will be bad (Table 8.1).

| Human: 94% Accurate | | |
|---|---|---|
| | Actually Good | Actually Bad |
| Tested Good | 46,060 | 60 |
| Tested Bad | 2,940 | 940 |

**Table 8.2** Human inspection performance.

With the human inspection, 6% of all inspections are incorrect, totaling 3,000 incorrect results, distributed as 2,940 false-negatives and 60 false-positives (Table 8.2).

| Vision System: 96% Accurate | | |
|---|---|---|
| | Actually Good | Actually Bad |
| Tested Good | 47,040 | 40 |
| Tested Bad | 1,960 | 960 |

**Table 8.3** Vision system performance.

The automated system only has a 4% failure rate, resulting in 2,000 incorrect results: 1,960 false-negatives and 40 false positives (Table 8.3).

| | Type 1 Cost | Type 2 Cost | Total Cost |
|---|---|---|---|
| Human | $ 300,000.00 | $ 1,470,000.00 | $ 1,770,000.00 |
| Vision | $ 200,000.00 | $ 980,000.00 | $ 1,180,000.00 |
| Difference | $ 100,000.00 | $ 490,000.00 | $ 590,000.00 |

**Table 8.4** Comparison of human and vision systems.

The overall cost of quality of the human system is $1.77 million, with the vision system costing $1.18 million, meaning the financial value of the information in this situation is $590,000 (see Table 8.4). This is a very simplified example, but it illustrates how to quantify the value of better information. When completing cost calculations, it is best practice to use benchmarks and actual data in favor of assumptions.

### *The Value of Faster Information*

The value of faster information comes from reducing the cost of delay in receiving information. To analyze how much information delay is costing, it is critical to understand what happens when information is delayed. In some cases, nothing may happen; a discrete part may be made with a defect that doesn't affect future parts being produced on the line, or a production system may have an inherent pause anyway so the delay in the information doesn't hinder the action driven by that information.

If each opportunity to create a defect is discrete and totally independent, then getting information at any time doesn't necessarily inform how the process is performing overall. This is not the case in many processes, however; information regarding the result of one product may drive an adjustment to the process because that process may drift and impact all future products. The cost of extra processing may also exist if a defective unit advances into a downstream process to be processed wastefully, and this may also create extra rework within the system.

Generally speaking, the main costs to look out for with speed of information are the delay in adjusting a process, the rework and quarantining activities that may be caused by this delay, and the extra processes that may be executed on a defective product. These costs can be further compounded by reduced throughput through yield loss, but a production facility should be designed to meet takt time based on customer demand, with quality rates included in the design. Improvements that ensure delivery on customer commitments should be used as a selling point for the ROI where they can be demonstrated; however, starting the analysis with the direct benefits defined above is appropriate.

Take a theoretical example where a production line has a takt time of one minute with all operations being balanced to a cycle time of one minute (this is a theoretical example to make the math easier). We will consider a human inspection that samples a product after an operation and takes 10 minutes to input results of analytical testing into the data system. This time is due to the logistics of collecting the parts to sample, executing the testing defined in the control plan, and inputting results into the database. The Quality 4.0 system will be an inline system, which measures in real time as the production line is flowing, so testing lead time for the inline system can be considered zero minutes. The example will use the following assumptions:

- The first pass yield of the production facility is 99%.
- It costs $10 per process step to process a part.
- This is a lean facility using single-piece flow with no work in progress (WIP) in between stations.
- When a defective part is found, the station must be maintained or more defects occur.

Determining the cost of the human system in this example is much easier. When the defect is produced and sampled, the line continues producing defects for ten minutes. The defective units make it through the next ten stations. The number of defects processed over the course of ten minutes, and the cost of processing, is demonstrated in Table 8.5.

| | Workstations | | | | | | | | | | |
|---|---|---|---|---|---|---|---|---|---|---|---|
| | a | b | c | d | e | f | g | h | i | j | Total |
| Minute 1 | 1 | | | | | | | | | | 1 |
| Minute 2 | 1 | 1 | | | | | | | | | 2 |
| Minute 3 | 1 | 1 | 1 | | | | | | | | 3 |
| Minute 4 | 1 | 1 | 1 | 1 | | | | | | | 4 |
| Minute 5 | 1 | 1 | 1 | 1 | 1 | | | | | | 5 |
| Minute 6 | 1 | 1 | 1 | 1 | 1 | 1 | | | | | 6 |
| Minute 7 | 1 | 1 | 1 | 1 | 1 | 1 | 1 | | | | 7 |
| Minute 8 | 1 | 1 | 1 | 1 | 1 | 1 | 1 | 1 | | | 8 |
| Minute 9 | 1 | 1 | 1 | 1 | 1 | 1 | 1 | 1 | 1 | | 9 |
| Minute 10 | 1 | 1 | 1 | 1 | 1 | 1 | 1 | 1 | 1 | 1 | 10 |
| Totals | 10 | 9 | 8 | 7 | 6 | 5 | 4 | 3 | 2 | 1 | 55 |
| Cost | $100 | $90 | $80 | $70 | $60 | $50 | $40 | $30 | $20 | $10 | $ 550 |

**Table 8.5**: Defective parts processed during the inspection period.

So, for each defect produced, $550 of extra processing costs will be incurred due to the lag time. There are two other factors to consider: the reduction of capacity of the system due to wasting capacity processing defects, and the total annualized cost. There are two conditions under which the production system may be operating: either it is not at capacity or it is at capacity. A system that is not at capacity does not need to operate at its full capacity to meet customer demands. In this case there is no quantifiable cost due to lost capacity, only the extra processing cost already captured. To determine total cost, the equation is very simple:

**(1–FPY) × (CPD) × (APT) = CSI**

**Where:**

- FPY = First Pass Yield (Percent)
- CPD = Cost Per Defect
- APT = Annual Production Throughput
- CSI = Cost of Slow Information

If we plug 50,000 units back into the assumptions the numbers are:

**(.01) × ($550) × (50,000) = $275,000.**

A more complex situation occurs when the system is at capacity, meaning that the production system can sell more if it can make more. There is an opportunity cost associated with losing capacity in this case. That opportunity cost is the number of units that could have been sold that weren't, and it would only be possible to calculate in the context of production cost and revenue on a per part basis. In this case, so as not to also write a whole section on capacity analysis (which is a different topic), we will keep the assumption that the facility produces 50,000 units per year. The cost of extra processing ($275,000 calculated in the example) does not change with capacity loss; it still exists. The capacity loss is incurred because when the defect is discovered, defective parts have been processed through ten stations, thereby taking away time that could have been devoted to processing good parts.

In this case, the defective parts would be cleared out from stations A-J. In a single-piece flow environment, those stations feed each other. There is no extra inventory to plug right into those stations, so Station A immediately processes the next good part and feeds it to Station B. One minute later, Station B processes a part to go to Station C. This pattern continues until Station J has a new part ten minutes later. From Station A onward, every station loses ten units of processing time, meaning that the entire line would lose ten units of production. To calculate the total loss, simply calculate the total defects produced annually multiplied by 10 and subtract from the whole. One percent of all products produced in this example are defects, so the equation is:

**(DR) × (APT) × (LPD) = PLA**

Where:

- DR = Defect Rate
- APT = Annual Production Throughput
- LPD = Lost Production per Defect
- PLA = Production Lost Annually

Which, when calculated is:

**(.01) × (50,000) × (10) = 5,000**

So in this example, this production system has lost 5,000 sellable units of capacity, which is likely a much larger cost than the $275,000 of extra processing costs incurred. This example uses very round numbers to illustrate the concept; a true capacity analysis involving cycle times and WIP inventory will be much more complex, and there are many resources available regarding capacity analysis. It's important to understand how the speed of information can impact overall capacity as an element of cost worth analyzing for ROI purposes when cost justifying a Quality 4.0 system.

### *The Value of More Information*

The value of more information comes from reducing the risk of detecting a defect in real time. In many sampling plans, only a small percentage of total production is sampled. Sampling plans are optimized to reduce the overall risk of producing defects, but the risk is never zero. If there is a period between sampling, it is possible for a process to go out of control between samples and produce a number of defects before being caught at the next inspection. The cost of this type of information gap is easy to see. In an ideal situation, the volume of information covers 100% of possible defects; the production system may be stopped and defects may be corrected at the first instance of a defect. This prevents any rework or scrap costs incurred beyond a single unit failure. This also reduces capacity loss, just as with faster information. Additionally, the gap in information requires an investigation of parts produced between samples to adequately quarantine all defects, as it is unknown exactly when a defect occurred between the last known good sampling and the first defective sample detected.

Consider an ROI example: A system which collects data from 100% of parts can be compared against a system that has intermittent sampling in play. For this example, we can assume that a process can go out of control at any time and this type of system would be best suited for processes that must be brought back into control after going out of control (meaning that after the first defect, more defects will be produced). If we consider a system where a human inspector tests one part every hour, with a throughput of 100 parts per hour for easy math, the value of having more information can be easily modeled with a theoretical Quality 4.0 system that inspects 100% of all parts. The human inspection, when it detects a defect, would require the parts produced between the last good sample and the defective sample to be quarantined. The 100 quarantined parts would need to be inspected, which incurs the labor cost and capacity reduction related to pulling parts from the line (as detailed in the example of faster information) as well as the scrap or rework costs associated with having extra defects. If this process goes out of control and stays out of control, and defects occurrence is randomly distributed, on average 50 of the 100 quarantined parts would require rework or scrap. The cost associated would be:

**(SRC) × (TD) + (IL) × (PI) = CPD**

Where:

- SRC = Scrap / Rework Cost per part
- TD = Total Defects
- IL = Inspection Labor cost per part
- PI = Parts Inspected
- CPD = Cost Per Defect

If inspection costs $5 per quarantined part and the rework cost is $20 per part, the cost incurred would be:

**($20) × (50) + ($5) × (100) = $1500.**

It becomes clear that defect costs can add up quickly and can be compounded by costs associated with capacity losses.

Much like previous examples, this type of analysis requires knowledge and context of the production system, the capacity, the potential for reputation damage from shipping bad product, etc. This type of analysis should be performed by experts who understand the production system—and in determining a total cost, the defect rate and the potential for one defect to lead to further defects must both be brought into play.

### One Last Thought on the Value of Information

When calculating the value and cost of information, as well as the cost of poor quality inherent within a system, remember that these numbers really require the context and knowledge of how a business operates. Calculating ROI usually demands some business knowledge and participation from people who really understand the system. These examples serve as a structural guideline for how to think about information and the value of said information, but each ROI and business case will be specific to an individual project. Developing the skills required to effectively and accurately calculate the ROI of a Quality 4.0 system will greatly improve the overall Quality 4.0 outlook and performance achieved by a business.

## Redistributing Quality Resources for Quality 4.0 Value

Another major consideration for the ROI of a Quality 4.0 system is how quality resources can be deployed. There is no straightforward calculation for maximizing value of quality resources, but a thorough analysis may be able to further justify the cost of a system. When thinking about quality resources and their value, asking how the system will change the scope of a quality department and the roles of quality professionals is critical. First, consider whether any inspection or analysis that will be automated frees up human labor. The value of this labor is hard to quantify, as it is highly contextualized. Consider a quality department with a team of inspectors performing sampling and measurement of a production system. If the Quality 4.0 system takes over the sampling and measurement, the inspectors can be redeployed to respond to problem conditions detected by the system. They may be utilized to quarantine defects sooner, reducing capacity loss and the waste associated with processing defective parts. The value of this type of change depends on defect rates and costs. To maximize the value of information coming out of a Quality 4.0 system, this is exactly how labor should be rearranged.

Second, understand that a system that generates highly valuable information drives priority setting and project needs. This allows for quality improvements to happen optimally. The better quality of information available, the more a quality department can focus resources on the costliest problems impacting the business. A Quality 4.0 system should account for overall improved performance of the business at a holistic level for this reason, although this is also difficult to quantify and project. Thorough costing and financial analysis on overall performance can help to project these numbers and use them to justify the system, but this also requires expert judgement and subject matter expertise on the operation of the production system and business as they exist currently.

### Putting it all Together: The Quality 4.0 ROI Calculation

The ROI calculation for a Quality 4.0 system is necessary to advance the deployment of the system. Business leaders likely don't want to spend money on a system that is not expected to improve performance tangibly, and in the context of business performance, improvements are usually seen through the lens of financial performance improvements. There is a cost associated with information that is not high quality, that is delivered too slowly, or is insufficient.

When the data strategy is scoped, the cost of current information versus existing information should be calculated in the ways described in this chapter. Projected improvements from optimizing the allocation of quality department resources should also be included. This must be considered against the cost of the system as it's scoped when setting system parameters. The functionality of the system cannot be compromised, or the analysis would need to be repeated with a new data strategy in mind—so the cost of the system should be clear at this point. It includes the cost of the technologies, including data collection, data management, data analysis, and information delivery technologies. The training cost that will come with system deployment should also be considered if external paid training is to be purchased for new technologies. Calculating the ROI and business case will then depend on the particular ROI method used by the company. Many companies utilize a simple payback period calculation with a 12-month threshold, meaning they calculate the savings (or the cost of poor information that will be reduced) on a monthly basis, and divide the cost of the system by that number. If it is less than 12, the system can be purchased.

Generating a comprehensive ROI calculation is critical to driving buy-in from business leaders and decision makers who must approve the purchase of the hardware and software to be deployed in the Quality 4.0 system. When performing ROI calculations, it is important to include subject matter experts, consider the context of the business itself, and also not try to pursue an unreasonable level of detail. The expected value of the system is ultimately an assumption. It is the responsibility of the quality practitioner to ensure the assumption is as accurate as possible within a reasonable context of the business, but it is an assumption, nonetheless. For this reason, there is a level of depth and detail that is unlikely to be beneficial to pursue when performing the analysis. Best judgment should be used to determine how specific an analysis should be. If the calculation shows the system is or is not worth purchasing by a whole degree of magnitude, the math does not need to be specific. For example, if a system will cost 10-20 times as much as it will save annually, it is safe to assume that the system isn't worth it instead of specifically doing a deeper analysis to project that the system will cost exactly 16.8 times the expected annual savings. If the initial numbers seem close, deeper analysis becomes valuable.

The ROI is the starting point of getting leadership to endorse new projects and the purchase of technology. This isn't any different with Quality 4.0. A good data strategy leads to an accurate system scope; it also allows for better calculations of potential value to be returned by the system. Getting leadership buy-in to purchase Quality 4.0 system components is critical in all Quality 4.0 implementations.

# Chapter 9
## Putting Quality 4.0 into Practice

THE IMPLEMENTATION OF Industry 4.0 to achieve improved performance has been a topic of interest for recent research. The implementation model explored in Chapter 7 is one such example, as is the framework shown in Figure 9.1 with key implementation factors.

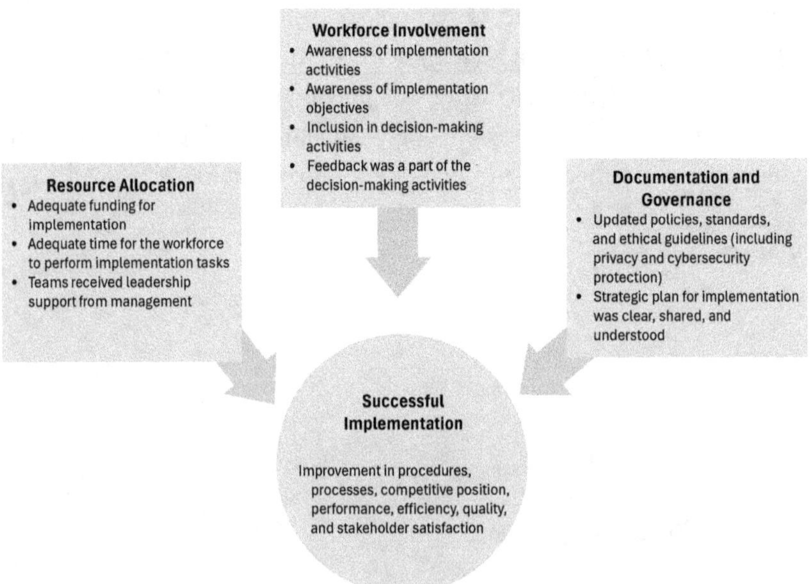


**Figure 9.1** Framework for critical success factors in Industry 4.0 implementation. Source: Khan, 2023.

This framework for success is layered over the top of the implementation model to create a high-level guideline on how to incorporate Quality 4.0. With a creative vision, this can be applied from a single project level to a companywide comprehensive deployment. It is important to note that the success factors and implementation models are not industry or application specific. This can also be seen in the presence of higher-level concepts, where more refined details and actions may need to be determined as appropriate for specific applications.

## The Critical Success Factors (CSFs)

In total, 19 CSFs were identified for Industry 4.0 success, as shown in Table 9.1, and bucketed into six categories for organizations to emphasize while planning and executing Quality 4.0 deployment.

| Category | CSF |
|---|---|
| Management | Management Commitment |
| | Management Training |
| | Communication |
| Workforce | Employee Involvement |
| | Employee Acceptance |
| Resources | Resource Allocation |
| | Technical Readiness |
| Strategy | Strategic Approach |
| | Governance |
| Sustainment | Performance |
| | Processes and Procedures |
| | Training and Education |
| Outcomes | Performance |
| | Efficiency |
| | Processes and Procedures |
| | Stakeholder Satisfaction |
| | Quality |
| | Competitive Position |
| | Success |

**Table 9.1** Critical success factors for implementation.
Source: Khan, 2023.

The listed success factors are well-articulated in the visual framework for practitioners to use.

### Resource Allocation

The organization must provide funding necessary to implement new technologies, while also ensuring there is sufficient time available to roll out said technologies. Throughout this process, the teams primarily responsible for implementation must receive support from leadership and their management structure to be successful.

### Workforce Involvement

Communication is a tool that must be used to empower the workforce while providing clear awareness of the objectives and activities related to implementation. Furthermore, because the workforce is considered users of these new technologies, their inclusion in decision-making activities is vital and allows for proper feedback throughout the decision-making process.

### Documentation and Governance

Policies, standards, and ethical guidelines must be updated by changes in technology and procedures to ensure success. It is also critical that a strategic implementation plan be clear, shared, and understood.

### Successful Implementation

The model for success is demonstrated in Figure 9.2.

**Figure 9.2** Successful implementation of Industry 4.0.

The key takeaway for successful implementation is that performance must tangibly improve along these topics, in alignment with the description of the data values shared in Chapter 8.

## The Model for Implementation

Revisiting the model for implementation, Figure 9.3 demonstrates the order of implementation activities presented in this book. Both the framework and the model have been validated through independent research projects, and it is worth noting that there are significant synergies between the two.

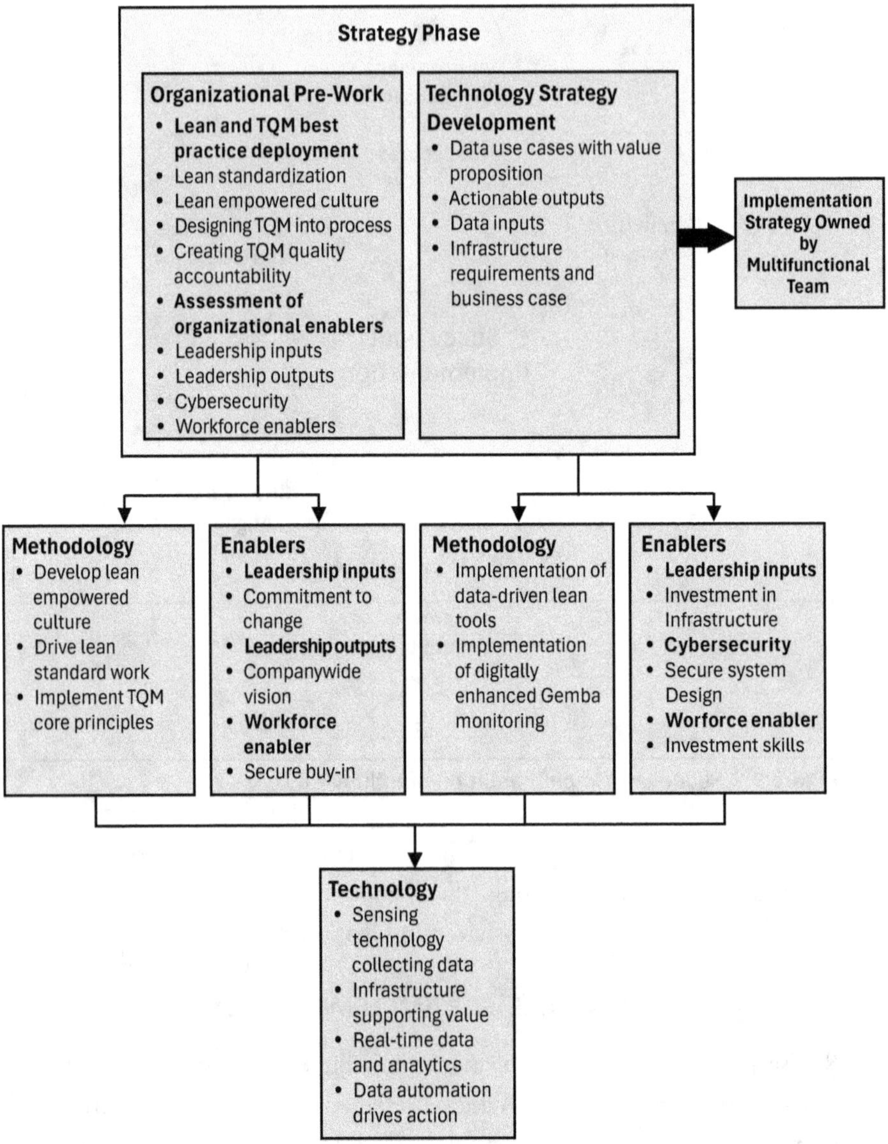


**Figure 9.3** Quality 4.0 implementation model.

## VALIDATING THE APPROACH

The CSF framework and implementation model were initially developed through reviewing the available literature on the topic. This literature review was used to understand the nature of how to implement Industry 4.0, as understood through the broader research community working on developing the newly identified capabilities that the fourth industrial revolution has empowered us to deploy. This first step is critical in deciding where to start when putting a new set of ideas into practice; however, it is also important to leverage real-world knowledge, experience, and examples. The model and framework both followed the approach of validating the literature findings by exploring knowledge available in the industry. Both research projects took the approach of extracting high-level ideas (described as codes and themes) from literature and validating them in the real world with appropriate data collection methods. The logic behind extracting codes and themes is in accounting for differences in terminology and application across different researchers and industries. The overall flow of how these approaches were validated is shown in Figure 9.4.

**Figure 9.4** Thematic analysis approach.
Source: Khan, 2023.

The framework of CSFs identified the aerospace industry as a source of information on real-world implementation of digital technologies. This was achieved through an industry survey of participants in the aerospace industry who had experience implementing digital technologies. The survey results were used to validate the final list of CSFs and examine their relationships. Similarly, the model for implementation was validated along the same approach of collecting the knowledge of individuals who have experience implementing Industry 4.0 technologies.

The implementation model utilized an expert interview approach, having deep-dive discussions with experienced professionals about the themes in the implementation model. This phased approach of compiling the most current research and validating it against real-world applications leads to both models being more practical to use.

One of the main challenges in writing about how to implement 4.0 technologies is the differences between industries and companies in their use of data, as well as their quality needs. In the aerospace industry, quality needs often have regulatory components. The industry utilizes the AS9100 standard, which is similar to the ISO 9001 standard, but with some key differences. Even within the same industry, companies will have different goals due to their businesses' competitive strategies and core competencies. All of this has to be considered when trying to develop a model for how to implement such a high-level idea.

What can be said with confidence is that the core ideas of both the framework and the implementation model have been validated through industry research. The next step in road mapping how to implement Quality 4.0 is to walk through an example highlighting how, when, and where to apply the concepts from the research.

## Walkthrough Example

This example describes the approach to implementing Quality 4.0 in a greenfield factory for a batch processing manufacturing process. The implementation example has been scrubbed of key proprietary information and focuses on the concepts of Industry 4.0 and improving quality performance, detailed generically in both supporting research papers.

### Background

The company that chose to deploy Quality 4.0 in this case had elected to roll out a new-build factory to accommodate future demand from multiple customers who were interested in their product at quantities beyond their current production capability. At the time of the project, there was significant growth for both the company as well as the overall industry, so there was an expectation that the capacity of the new build would be saturated and the company would need to follow up with future expansion in the near future. With this in mind, the goal was set to ensure this factory would perform at a high level and establish a roadmap for future growth projects. This goal would ensure customer confidence in the company's ability to deliver. The company produced bulk quantity commodity products that sold into the upstream supply chain of multiple other industries, which satisfied both end consumer product manufacturing as well as business-to-business (B2B) sales. As such, their production model was a batched production cycle that produced a small number of products with high volume.

The company elected to fully integrate digital systems to ensure quality product output by the factory, which led to the initiation of a Quality 4.0 project. (This developed organically; however, it is reasonable to expect that moving forward, more and more scaling and new product introduction projects across industries will organically result in companies wanting to integrate digital systems to ensure improved performance. In many cases, these projects may not be labeled "Quality 4.0," but projects that focus on integrating data usage to improve company or process performance will benefit from the success factors and model presented in this book.)

## Project Kickoff: Strategy Phase

The first step in architecting a successful digital system is the strategy phase. The strategy phase can have different scopes, depending on the scale of the project itself. Figure 9.5 highlights key points in the strategy phase of a Quality 4.0 project.

In this instance, the scope was a fully integrated facility. The goals for this project were laid out at the corporate strategic level and centered on the value proposition of data.

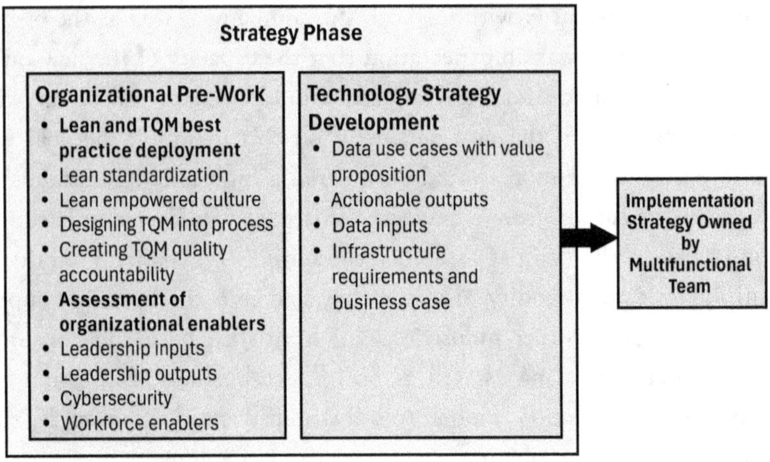


**Figure 9.5** Strategy phase.
Source: Jones, 2023.

The organization kicked off the initiative with a comprehensive workshop, including a cross-functional team. This was important, as it leaned into multiple identified CSFs, including:

- Employee involvement—getting different inputs from different departments and skillsets at the outset
- Employee acceptance—allowing employees to shape the strategy and thereby increasing buy-in for the project
- Management communication—collaborating with many functional groups to deliver the objectives and commitment clearly
- Strategic approach—putting tangible resources into developing a digital strategy leadership committed to this CSF
- Management commitment—committing to the effort by having the comprehensive strategy workshop management

The leadership of the organization was firmly committed to the technology strategy side of the strategic development phase of Quality 4.0 implementation. The workshop was designed to specifically map out how data would be utilized within the system. The first step was to identify the key performance targets that would cause the factory to be successful. In this case, the factory was already committed to run at full capacity. Operationally key challenges were already identified:

- Process steps produced large batches with long lead times.
- Quality testing had long lead times.
- Batches were blended in downstream process steps.
- Customers within the industry expected full batch traceability.

These key factors led to the organization setting goals of digital integration connecting with quality test data and batch records, in real-time, to prevent sending off-specification product downstream to contaminate a known "good" product. In addition, due to capacity demands, the system's goal was to continue to run without always waiting for quality test results. These goals were brought to the workshop to develop a digital strategy.

The components of a digital strategy include:

- Data use cases with value propositions
- Actionable outputs
- Data inputs
- Infrastructure requirements and business case

For this project, the organization fell into one of the common pitfalls seen in software integration, which is selecting software prior to architecting the process. Multiple software systems had already been scoped and selected by the organization leading up to the workshop, which included independent systems for:

- Enterprise resource planning (ERP)
- Material resource planning (MRP)
- Digital manufacturing integration

- Manufacturing execution system
- Quality test database
- Data and KPI reporting
- Maintenance scheduling and work order management
- Safety incident reporting

These software systems and their capabilities were brought to the workshop to architect how they would be utilized together to achieve the companies' goals; however, this is the wrong approach. The correct first step when laying out software architecture for an integrated digital system is to map out the system and how it needs to function. This will ensure proper performance.

In the workshop, the process was ultimately salvaged as the company had brought in two independent consulting groups with experience in digital transformation and had also brought on an internal practitioner with knowledge of the implementation model (referenced earlier in this chapter). This led to the workshop being a comprehensive effort to map the tangible processes of the factory and then map the corresponding digital integration that was required for goal achievement. By bringing in the correct resources needed to get the project aligned with the goals and vision, the company further leaned into both management commitment and resource allocation CSFs.

## Technology Strategy Development

The workshop was a good first step, but it would not have been useful if it didn't realize tangible results towards the goals. It was important to know the goals for the factory going into the workshop so the strategy could be effective in meeting them. The tangible outcomes from the workshop included data use cases with value propositions.

The team elected to fully integrate measurable process data on all machine parameters at every batch step. The value proposition was to have a leading indicator on process performance, since quality testing had a long lead time. This would be similar to measuring the operating temperature of a cutting tool in a machine shop. If it is impossible to know if a surface finish is acceptable until the cutting is complete, the temperature of the cutting tool may indicate when the process is going off-spec prior to the part being damaged.

The cross-functional team identified several machine parameters to not only monitor, but also to automate into a single data reporting system with alarm functionality to prevent potentially bad batches of product from being moved forward in the production system.

A second data use case was to fully integrate the MRP material transactions with the quality test results of completed batches. This would automate the reporting that satisfied the customer traceability requirements. This particular data use case had the value proposition of time savings, as manually generating these reports was time-consuming and included the potential for human data entry errors.

The traceability and process performance value propositions were the primary data use-cases of the system, as they satisfied the main goals the company had for the factory. Other secondary data use-cases were mapped as well, but for readability, this example will focus on the process control aspect of the system as the most important goal.

The actionable outputs for both identified value propositions were very simple. As a general rule, the actions *should* be simple, as the main reason to create a complex digital architecture is to accelerate analysis and get to straightforward decision-making. Any decision that leads to action which is already well understood and won't benefit from supporting data collection isn't a great candidate for a Quality 4.0 project; and if the information generated by a Quality 4.0 system is so complex that it doesn't support a straightforward decision, then the project itself could be called a failure.

Remember the data > action relationship detailed in Chapter 4:

1. Select an action that impacts the value proposition.
2. Scope the decision that drives the action.
3. Define the information that best feeds the decision.

4. Design the analysis that creates the information.
5. Identify the data that feed the analysis.

For the process monitoring and product quality value proposition:

1. The action is to quarantine or advance a batch of material.
2. The decision is whether the process is operated correctly or not.
3. The information is if the operating conditions are within quality performance windows.
4. The analysis is a combination of R&D generated knowledge of performance windows and continuous analysis of product quality data and process performance on an ongoing basis.
5. The data are the physical properties of how the equipment operated (such as temperature).

In this case, everything leading to the physical quarantine was automated. The system would not automatically advance material; a human had to interface with the controls to send material downstream, so the actual action towards the value proposition was in choosing to advance the material or trigger a manual process to remove it from the normal process flow.

All of those data > action relationships were mapped within the workshop, including in-depth decisions about the physical properties to be monitored. At this point, the knowledge in the workshop was used to determine which properties could and should be measured; however, specific sensors and technologies were not included yet, as those are a downstream requirement. It is important to avoid the pitfall of mapping an impossible system, and for this project, having individuals with knowledge of industry standards and sensor use-cases was crucial. If a Quality 4.0 technology strategy development event does not include individuals with knowledge of what data can be collected, then the project is being executed without the right people.

## Organizational Pre-Work

In this workshop, most participants were focused on the technology strategy, which is not strictly a problem; however, the organizational pre-work must also be considered. Quality 4.0 success correlates heavily with two key methodologies, which are not new and should not be reinvented:

- Lean manufacturing
  - Culture of empowerment
  - Standardization
- Total quality management (TQM)

These factors are critical to Quality 4.0 success because a Quality 4.0 system, when deployed successfully, generates easy-to-follow information that should lead to better actions. This is impossible to achieve if people aren't empowered to act (lean culture), accountable for quality outcomes (TQM), and clear in what their action and response plans are (standardization). Implementing these aspects of the strategy phase is not as technical as mapping the data plan, but lean and TQM should be considered and deployed nonetheless. As a bonus, an organization can begin to roll them out before any new technology is unveiled.

In addition to lean and TQM, the implementation model dictates an assessment of enablers, which aligns with the CSFs presented in the framework. These activities should be rolled out concurrently with the technology strategy, and if Quality 4.0 is the goal, lean and TQM should be rolled out immediately, as they are methodologies that don't require technology spending. This process can start with the organizational assessment of CSFs. No formal audit has been developed for this assessment; however, using a skilled auditor, especially one who has worked with external standards such as ISO 9001, would be a good place to start.

In the context of the use case, the Quality 4.0 practitioner within the company assessed the gaps and initiated company-wide lean training, as well as an effort to develop standard work for the critical processes in the upstream R&D facility. This effort was designed to familiarize the organization with standard work and lean principles to increase overall maturity, leading into the new facility in development. The assessment of the organization found that the core principle of TQM, in which the quality of the product was the responsibility of everyone within the organization, was already foundational to the process owners and subject matter experts responsible for starting up new processes. (If this had been a gap, TQM would have had to be deployed at this time as well.) Schedule-wise, the workshop took place approximately 18 months before the factory was expected to start putting out product. It should have been enough time to lay the groundwork for these methods, but commitment from the organization and leadership was necessary to do so.

The lean deployment plan involved organization-wide training on waste and value identification, followed by implementation of Gemba walks as a cultural empowerment tool. A 5S (sort, set in order, shine, standardize, and sustain) method was also deployed to drive standardization from the ground up. The second phase of the company's lean transformation was to improve standard work documentation and begin training A3 (the process problem solving tool) thinking to set the culture of continuous improvement.

When creating a rollout plan for either lean or TQM, it is important to remember that there is a wealth of knowledge available on these topics, and they do not need to be reinvented to be useful for Quality 4.0. These methods should be introduced according to standard best practices.

Assessing the organization at the strategy phase focuses heavily on the input side of the CSFs, specifically the ones shown in Table 9.2. These CSFs can all be established at the outset of a Quality 4.0 deployment, with Sustainment and Outcome-based CSFs being more important at the end of the effort.

| Category | CSF |
|---|---|
| Management | Management Commitment |
| | Management Training |
| | Communication |
| Workforce | Employee Involvement |
| | Employee Acceptance |
| Resources | Resource Allocation |
| | Technical Readiness |
| Strategy | Strategic Approach |
| | Governance |

**Table 9.2** Strategy phase CSFs.
Source: Khan, 2023.

Many of these CSFs were addressed through the workshop hosted by the senior leadership of the company. Two key gaps from the workshop were the Technical Readiness and Governance CSFs, as the company had allocated resources for software systems but had not committed to external training on the usage of said systems (a critical step in developing the Governance CSF). Some aspects of Governance were being targeted through the lean deployment effort, as standard work documentation was maturing and had the potential to naturally grow into well-documented Governance practices.

Not committing to the training for new software systems is another major pitfall within a Quality 4.0 implementation.

The final specific strategy component to be considered was cybersecurity. Digital integration always poses a cybersecurity risk. Cybersecurity skills and systems must be rolled out in the same way that other software and digital technologies are integrated in a Quality 4.0 implementation; however, it cannot be stressed enough that cybersecurity should be specifically assessed. During the workshop, a systems specialist internal to the company specifically audited the cybersecurity robustness at the company's headquarters and uncovered significant gaps, which led to a focused effort to both shore up current systems and a focus on architecting the necessary cybersecurity into the future state system.

***Strategy Development Steps for Practitioners***

The step-by-step approach taken in the workshop, in addition to the organizational assessments performed by specific individuals, highlights the method for developing a Quality 4.0 strategy. In this case, it is worth noting that the company had already selected software applications for purchase, and in many cases, when attempting to implement Quality 4.0 in a legacy process, a practitioner will have to work with the software systems they have available. In the case that strategy is developed with a clean slate on software applications, software down selection should take place after the technology strategy is written to ensure the purchase of software that meets the needs of the process.

Implementation model steps to follow:

1. Clarify quality goals.
2. Map physical processes.
3. Identify data use case value propositions.
4. List actionable outputs.
5. Identify data inputs.
6. Architect data infrastructure.
7. Down select and purchase software.
8. Comprehensively train users.

All of these steps were followed or planned through the digital architecture workshop put on by the company. A critical win highlighted by this process was the comprehensive inclusion of operational business units, as opposed to having an engineer or IT professional lead a focused project from only their perspective. The team in this case was able to identify a value proposition related to how data would be used at the new facility, which was the cost avoidance of allowing off-spec material to move through the facility. This was well-aligned with the company vision of being a high-quality producer of highly technical engineered materials. One thing to note is that Quality 4.0 does not require a new approach to strategy development in general; it simply adds the data and analysis aspect into legacy approaches. When rolling out Quality 4.0, companies will achieve greater success by using as many standard approaches as possible and only creating brand-new methods where necessary (as is the case with the lean and TQM deployments).

### CSFs to Leverage

The company workshop provides a clear example of leveraging CSFs for Quality 4.0 success. Table 9.1 lists the primary success factors to highlight at the strategy phase, as discussed below:

- **Management Commitment.** Implementing a comprehensive technology package will always be a larger project than is expected by first-time practitioners; to be successful in this case, the management team committed the time and cross-functional resources of the company. They also brought in external resources to leverage experience and expertise that was not available within their staffing and talent pool.
- **Management Training.** The leadership in this company set the goals of the project themselves and then deferred to the external experts on how to go about achieving them, including themselves in software trainings to learn the capabilities and limitations of the systems they were bringing in.
- **Communication.** At the outset of the workshop, the senior leadership team clarified the goals to the subject matter experts in the company, clearly communicating

the vision for how the future factor would be operated and the roles those experts had to play in building a system that would drive ease of use and correct decision-making for downstream employees.

- **Employee Involvement.** The company committed to employee involvement from the outset of the project scoping effort, as demonstrated by the commitment to the cross-functional workshop approach to scoping the value proposition and data architecture.
- **Employee Acceptance.** This was not guaranteed through the workshop effort; however, to ensure employee acceptance, typical strategies for buy-in are the most successful approach. Empowering different functional groups to design the system to suit their needs did drive the buy-in that was needed in this case.
- **Resource Allocation.** In this case, things were done out of order. The company had already invested heavily in software packages and chose to follow up with implementation consultants and an employee-centered workshop to ensure the success of said investment. In an ideal Quality 4.0 deployment, digital strategy would be architected and then software would be sourced to suit the needs of the project. Resource shortages often occur when companies balk at technology investment and end up trying to piece together an incomplete system when they have cut out critical pieces of technology, such as back-end software or sensor packages. In this case, the company invested fully in the technology package from the outset, allocating sufficient resources for the system.
- **Technical Readiness.** This success factor was demonstrated in this case in two ways. First, the use of external consultants who specialized in the implementation of

digital systems was critical to scoping the interconnected software/hardware architecture in technically feasible ways. Second, the company selected employees to receive external training in the software systems that would require direct employee interaction to ensure that the system would not fall apart as soon as the software vendors handed it over to the company.

- **Strategic Approach.** The goals and commitment to the digital transformation workshop were centered around taking a strategic approach to the implementation of purchased software systems.
- **Governance.** In this case, governance was not fully established at the time the workshop was completed. The company was growing into a new facility with a digitally integrated vision, and software governance was not already part of the existing systems, as they were largely much more manual. Taking the strategic approach to defining the use cases and goals of different systems, as well as their requisite data, laid the foundation for a methodical approach to software governance in the future. This is a CSF that is necessary for Quality 4.0 success, as it is a common pitfall for a company to attempt to maximize the usage of an expensive software package—even if it dictates supplying data that isn't beneficial to the mission of the company. Remaining goal-oriented was the key element of the governance CSF.

Pitfalls to avoid:

- Putting software before the process
- Overcomplicating the decision point of the data > action cycle

- Not including subject matter experts (internal or external)
- Not committing to training for new software and systems
- Assigning digital transformation as a project to the IT team

This case provided a comprehensive example of how to develop a digital strategy. Overall, the only element of the company's digital workshop that should have been changed was the decision to purchase software before the development of a digital strategy—although it is feasible to consider that the sunk cost invested in digital systems was a motivating factor for resource allocation and leadership commitment to the project.

## Software Down Selection

For practitioners who choose to develop a strategy before purchasing software, the process to follow is software down selection. The goal is to identify software that suits the data needs to achieve the value proposition. Down selection is the process of elimination that comes from identifying which software may not suit the needs of the process. Having a well-architected data strategy is critical to ensuring that a company purchases digital assets that fill all the needs of that strategy. A down selection process may be executed as follows:

1. A Quality 4.0 expert identifies the need for automated temperature data collection every minute, with an alarm being sent by text message to a maintenance professional if the temperature increases beyond a set threshold. The value proposition is the maintenance professional responds to an out-of-control machine condition as quickly as possible.
2. The system is scoped to do the following:
    a. A sensor constantly monitors temperature.
    b. A data point is recorded every five minutes as a rolling average of the previous five minutes.

    c. An algorithm compares data points to a known control limit.
    d. Automated communication is sent wirelessly when the algorithm detects failure.
3. The Quality 4.0 expert considers five software packages that offer the ability to run the temperature alarming algorithm.
    a. One is eliminated because it lacks automated communication functionality.
    b. One is eliminated because it operates in the cloud and takes too long to transmit data to and from the site.
4. The remaining three choices are considered for the following criteria:
    a. Cost to purchase and maintain
    b. Technical support for installing and implementing
    c. Training offerings for company employees to become subject matter experts

This is a simplistic version of a down-selection process. Like many capital purchases, a set of criteria must be used to make a final determination; however, if the down selection process fails to eliminate the first two options against the requirements of the system, the company may end up with software which can't adequately fulfill the needs of their digital strategy so they would fail to realize value from their data.

## Total Quality Management and Lean Deployment

The other aspect of the strategy phase is developing the organization for Quality 4.0 success. Lean manufacturing and TQM are tried-and-true manufacturing practices that both contribute heavily to Quality 4.0 success.

Lean manufacturing centers on developing a culture of continuous improvement to achieve a production system that works just in time, based on customer demand, and creating a pull. This requires the precise coordination of several functions and processes, which is not achievable through brute force of system design. Achieving this level of coordinated precision requires incremental improvements across the entire system. To get all the way to a fully mature, single-piece flow requires a culture of continuous improvement. This culture of continuous improvement can be empowered with the deployment of a digital system that provides accurate information to accelerate or improve (or both) decisions made in real time. If a factory has a culture in which everyone is expected to contribute to the improvement of the system, empowering everyone with information will set them up for success. This ties to the empowerment aspect of realizing value from data, which is described in Chapter 8.

Total Quality Management coordinates with Quality 4.0 largely for the same reason. In a TQM-based quality system, quality becomes accountable to every individual, by design. If a workforce is expected to maintain the quality of the product and process steps they work on, information that shows clear, real-time status of the quality of that system helps enable the success of those employees. The information and data relationships support both the empowered culture and the accountable culture of these two methods; it can be stated that Quality 4.0 systems successfully combine lean and TQM best practices for this reason.

### Lean Deployment

Lean manufacturing has been written about in countless works and has been celebrated around the world as highly successful in driving manufacturing profit and performance. Therefore, a Quality 4.0 professional should avoid reinventing the wheel when it comes to deploying lean. Many companies may already be using certain lean tools before their Quality 4.0 journey, such as 5S or Gemba walks, and some companies may be fully mature with lean manufacturing before even deciding to go through a Quality 4.0 deployment. Regardless of a company's comfort with lean manufacturing, it's important to create a strategy for implementing an empowered lean culture.

The Quality 4.0 expert in the case being explored chose to deploy two primary lean tools to prepare for the future digital architecture: 5S and Gemba walks. 5S is an organizational tool designed to standardize workspaces. Many companies use 5S principles without intentionally using lean manufacturing because being organized has many benefits when it comes to completing standard work. A surgeon, for example, may have never heard the term 5S, but they have a very detailed layout of how surgical instruments are presented and staged during their procedures. This is an example of 5S. 5S is often the very first tool rolled out when deploying lean, as it is very straightforward and sets the foundation for standardized work by creating a standardized workspace. This first layer of standardization also provides benefit to the future Quality 4.0 system, as receiving standard information to make decisions is challenging within an environment that isn't standardized at all.

5S is also a very important tool to drive organizational and leadership buy-in to lean manufacturing methods. It sets a foundation that is visually appealing, and the implementation of 5S is very tangible to anyone who walks into an organized facility—especially if it was once disorganized and the improvement can be felt. These reasons are why 5S was the first tool deployed by the Quality 4.0 practitioner. Many resources detail how to roll out 5S within a facility.

Gemba walks were also selected to prep the organization for Quality 4.0, because Gemba walks are one of the most powerful cultural enabling tools in lean. Gemba means "the real place" in Japanese, and Gemba walks are simply walks by leaders on the shop floor, or the real place where customer value is created. The purpose of a Gemba walk is to facilitate face-to-face discussion between leaders who supply resources and employees who encounter daily problems. The idea is to drive problem solving, continuous improvement, and buy-in from value-adding direct labor. The Quality 4.0 expert in the case began the Gemba process by training the entire staff on value and waste, teaching everyone to recognize different lean wastes, which are summarized in the acronym DOWNTIME:

- Defects
- Overproduction
- Waiting
- Not utilizing talent
- Transportation
- Inventory
- Motion
- Extra processes

This training set the foundation for direct labor employees to be able to articulate the challenges they faced to leaders in a standard language. The second step was setting up Gemba boards and bringing leadership of various functions and levels on regular, scheduled Gemba walks. Further details on Gemba board design and Gemba deployment can be found in several resources.

After 5S and Gemba, many different paths may be taken, and it may be worthwhile to develop employees with specialized lean training. These efforts to develop an empowered culture are critical to long-term Quality 4.0 success.

## Total Quality Management Deployment

Total Quality Management is another method that has been long established in manufacturing. The critical element of TQM that pushes Quality 4.0 success is that everyone is accountable for the quality of the company. In the case at hand, the company culture was driven by a technically advanced, highly motivated core team in developing high-tech materials for an industry that was largely valued in the eyes of the company's employees. A culture of quality accountability was a natural byproduct, and with the relative youth of the company, quality had been made every employee's job out of necessity due to the lack of an established quality department with fully mature processes for quality control and quality assurance.

The American Society for Quality (ASQ) outlines several elements of TQM. The method is:

- Customer-focused
- Employee involved
- Process-focused
- About continuous improvement
- A systematic approach
- Fact-based

All these facets of an approach to quality coordinate well with the purpose and scope of a Quality 4.0 system. When a Quality 4.0 practitioner must deploy TQM, creating a plan to roll out these aspects of a quality system and using available implementation resources are critical to success.

## Lean and TQM CSFs

Lean and TQM both leverage many of the same CSFs as the strategy development phase of the project; however, both methodologies on their own will also begin to provide outcomes as well.

| Category | CSF |
|---|---|
| Management | Management Commitment |
| | Management Training |
| | Communication |
| Workforce | Employee Involvement |
| | Employee Acceptance |
| Resources | Resource Allocation |
| | Technical Readiness |
| Strategy | Strategic Approach |
| | Governance |

**Table 9.3** CSFs revisited.
Source: Khan, 2023.

| Category | CSF |
|---|---|
| Sustainment | Performance |
| | Processes and Procedures |
| | Training and Education |
| Outcomes | Performance |
| | Efficiency |
| | Processes and Procedures |
| | Stakeholder Satisfaction |
| | Quality |
| | Competitive Position |
| | Success |

**Table 9.3** CSFs revisited. (continued)

Lean and TQM both leverage all the outcome-based CSFs when fully deployed (Table 9.3). At a high level, lean, TQM, and Quality 4.0 work together to realize value in manufacturing through the day-to-day decisions and actions made by an entire workforce. Empowering individuals to act and make small improvements, driving accountability towards quality and performance throughout an entire workforce, and putting the right information in the hands of everyone to make the right decisions at the right time are three different ways to describe an effective workforce. Success in all these methods requires levels of engagement, collaboration, trust, empowerment, and a strong culture.

## Post Strategy Phase

At this point, the Quality 4.0 expert was well into the middle phase of the implementation model (Figure 9.6).

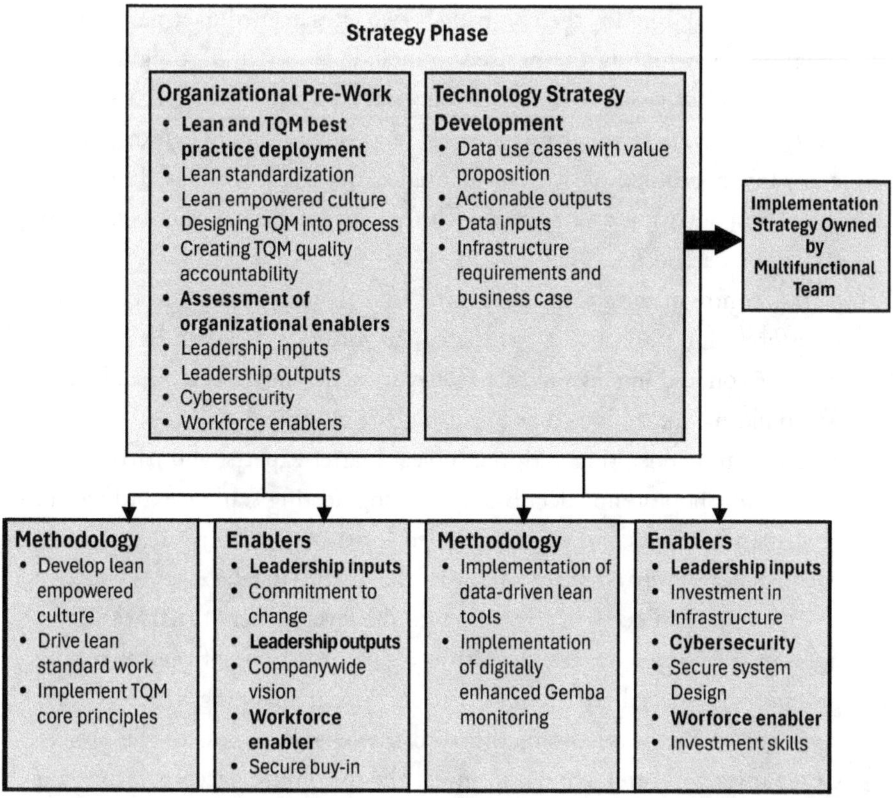


**Figure 9.6** Quality 4.0 implementation model.
Source: Jones, 2023.

In this case, the actions of the middle of the model have largely been described as they were. This section of the model is not reflective of bringing in new tasks; it is about the continuous effort and time required to deploy these methods and tools. In the example of 5S, the company first had to approve the 5S program in concept and detail. This took approximately one month, which is normal for any corporate process being deployed. Then training materials were written and distributed, trainers were trained, and staff were trained. This effort took several more weeks across multiple sites. After training, red tag events had to be scheduled by area, which again took weeks. Overall, deploying 5S across the company took approximately six months. Throughout this time there were shifting levels of priority, firefighting, and iteration on what

5S was to look like for the company. (Two sites argued over naming the program 6S, which is a term used to include Safety in 5S programs.)

This level of sustained effort is necessary for all Quality 4.0 support work. While all of this was going on, and day-to-day operations at the company continued as well, a focused technical team was developing the actual software and system architecture detailed in the technology strategy workshop.

It is common for a team of IT professionals to take the lead on digital transformation projects such as these, and for the systems to be deployed in the vision and interests of that IT department. In this case, the IT team was building the architecture but using the output of the workshop as a roadmap and consulting with the subject matter experts who participated in the workshop on fine details. Support engineering staff were brought in to shape the inputs and outputs needed for the MRP software system to adequately receive and share data with the quality database.

One critical success in this case was the level of detail that was developed in the workshop, largely with the support of one of the consulting groups who specialized in information system mapping. These efforts leveraged at the outset were critical to the overall success of the project. Companies will often choose to invest the minimum acceptable amount into quality success at the beginning of a major project or campaign; however, the relationship between early spending and total spending on quality is typically such that spending early avoids future costs and saves money in the long run. This relationship is visualized in Figure 9.7.

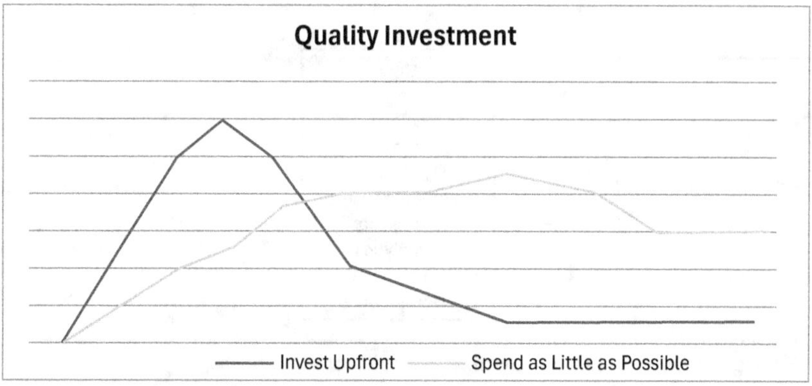


**Figure 9.7** Quality spending strategies.

## Finalizing Quality 4.0 Deployment

At the time of this writing, the value propositions established in the workshop were being validated, but with promising early success. The goal of this system was to identify and quarantine off-spec product as early as possible to prevent contamination and loss of capacity at the new facility. Software systems were well coordinated, largely due to highly detailed planning at the outset within the workshop. A successful Quality 4.0 deployment centers on detailed architecture built to realize value from data, followed by a step-by-step road map explaining how to roll out the system, which will act out those relationships. Figure 9.8 shows the final phase of Quality 4.0 implementation. If the preceding work was thorough and effective, this phase is as simple as installing sensors and software, training staff, and validating the system when it's turned on.

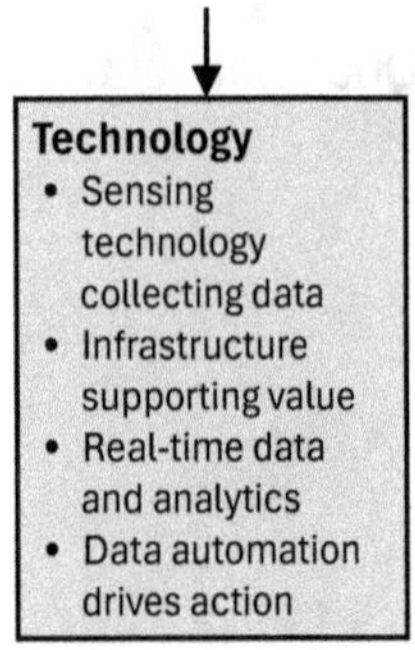


**Figure 9.8** Quality 4.0 deployment completion.
Source: Jones, 2023.

### Putting It into Practice

The example in this chapter is generalized at a high level because no two systems will be any more alike than two companies' quality goals and policies. It is critical to understand how to define and architect a value proposition from data, followed by scoping the technical system requirements to achieve said value proposition. Everything in this chapter is based on industry research and real-world implementation, with key points in deploying Quality 4.0.

- Identify, define, and leverage CSFs.
- Clarify value propositions from data.
- Develop a clear digital strategy to realize data.
- Leverage existing lean manufacturing methods.
- Leverage existing TQM methods.

## Other Real-World Examples

As a final snapshot of the direction Quality 4.0 is heading, there are examples that can be referenced for current success in articles and online. The implementation of Quality 4.0 in industries refers to the integration of its tools in the prevention of defects, improving efficiencies, innovation and creation of smart products, processes, and frameworks to delight customers and increase profitability in industries. Here are some success stories for the implementation of technologies such as automated ERP systems, IoT, machine learning, and AI. These efforts are intended to integrate Quality 4.0 and its innovative technologies.

### Implementing Quality 4.0 in Higher Education[2]

In his research for developing a framework for assessing the maturity level of implementing Quality 4.0 in higher education institutions, Alzahrani (2022) stated that "the application of Quality 4.0 practices in higher education institutions (HEIs) has the potential of driving innovation through technological advances in various aspects." His research "aims to develop a framework to assess the maturity level of HEIs towards transformation to Quality 4.0 while considering the impact of such transformation on the process, people, and technology dimensions defined in the Quality 4.0 model proposed by the LNS Research."

Alzahrani's literature review revealed that HEIs used big data for different applications that include decision making, reporting and compliance, analysis and visualization, security and risk mitigation, predictive analytics, data generation and acquisition integration, data warehousing, and performance evaluation. The proposed framework has been validated by determining the maturity level of Quality 4.0 implementation within Saudi HEIs. The Friedman test was performed to statistically ensure the framework's ability to differentiate between the maturity levels of adopting Quality 4.0 observed in the study sample. The result showed that all 26 Saudi HEIs obtained scores ranging from the second to the fifth levels of the Q4.0-MS." This finding indicates that some HEIs have a basic understanding of digital transformation, while others are at the most advanced level of Quality 4.0 adoption (Figure 9.9). The performed analysis confirmed the applicability of the developed framework for assessing the level of adoption of Quality 4.0 axes implementation at any population of HEIs.

2 Bandar Alzahrani's "A Framework for Assessing the Maturity Level of Implementing Quality 4.0 in Higher Education Institutions," Ph.D. dissertation, University of Central Florida. (2022)

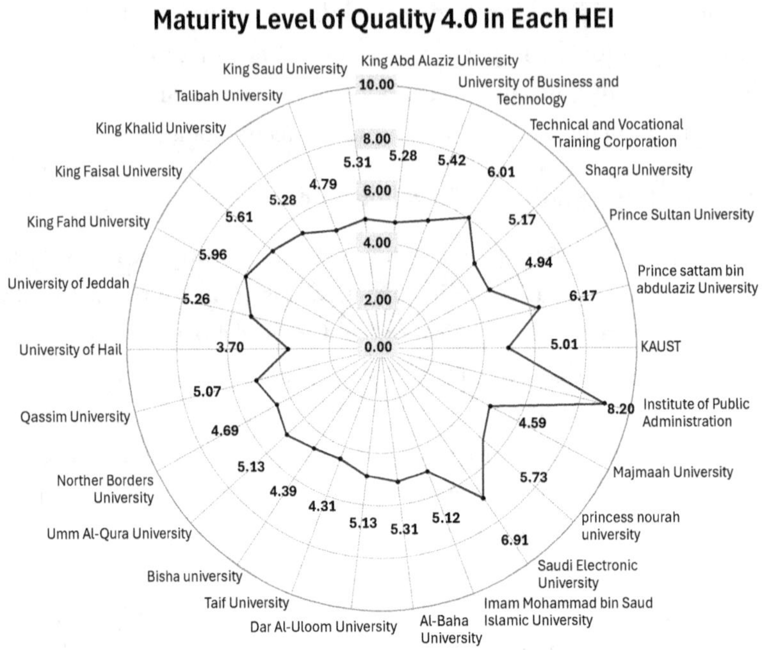


**Figure 9.9** Maturity level of Quality 4.0 in each HEI in Saudi Arabia.
Source: Alzahrani (2022).

## AI-Driven Predictive Maintenance in Manufacturing[3]

Incorporating AI in predictive maintenance offers some benefits and outlines some challenges that must be overcome. Among these challenges are data security, data integration, machine readiness, cost, and expertise. The integration of AI and predictive maintenance also comes with more benefits and gains that include enhanced efficiency and downtime reduction, cost efficiency, and enhanced safety.

AI-driven predictive maintenance in manufacturing is increasingly being recognized for its potential to significantly reduce downtime, improve equipment efficiency, and contribute to substantial cost savings. Companies across various sectors are leveraging AI to predict equipment

3 "AI-Driven Predictive Maintenance in Manufacturing" by Rishika Patel, February 12, 2024. Downloaded January 5, 2025 from: https://cioinfluence.com/it-and-devops/ai-driven-predictive-maintenance-in-manufacturing/.

failures before they happen, enabling proactive maintenance that keeps operations running smoothly (Patel, 2024).

1. **Rolls-Royce:** They use IBM Maximo and Watson to monitor their jet engines in real time, predicting potential issues before they arise. This has led to a 25% reduction in unplanned downtime and a 10% increase in engine life.
2. **GE Aviation:** They use GE Predix, their own AI platform, to predict maintenance needs for their jet engines and other aircraft components. This has reduced unplanned downtime by 50% and saved them over $1 billion in maintenance costs.
3. **Schneider Electric:** They use an AI system called EcoStruxure Predictive to monitor their electrical distribution equipment. This system has helped them reduce unplanned downtime by 30% and improve operational efficiency by 10%.
4. **United Technologies Corporation (UTC):** They use a platform called Predix from GE Digital to monitor the health of their aircraft engines and other components. This system has helped them reduce maintenance costs by 20% and improve on-time delivery rates by 5%.
5. **Ford Motor Company:** They use an AI system called Predictive Engineering Analytics (PEA) to predict potential failures in their vehicles. This system has helped them identify over 100 potential failures before they occur, preventing costly recalls and safety issues.

## Leveraging Artificial Intelligence in the Laboratory[4]

Lab Leaders (2023) stated that "Laboratory personnel utilizing the power of artificial intelligence (AI) can expect to see many transformative benefits. Whether taking over the performance of repetitive and mundane tasks or easing the workload of personnel in understaffed laboratories, artificial intelligence (AI) is a tool that is reshaping the ways in which laboratory personnel perform their work." Among the different benefits that the article listed are:

1. Managing laboratory operations so personnel don't have to
2. Advanced image analysis to support rapid and accurate diagnosis
3. Making personalized medicine the standard of healthcare
4. Capturing actionable insights from big data
5. Pushing the needle in drug discovery

However, the article (Lab Leaders, 2023) listed some barriers to the implementation of AI in the laboratory, that include:

1. It will be crucial to ensure that AI-based algorithms are trained using representative and unbiased data. Such mistrust of AI is a barrier to its use. AI systems must therefore be transparent and explainable.
2. There is also a common belief that AI will result in job displacement. This must be tackled to ensure AI is seen as a tool that can enhance human capabilities rather than replace them.

4 "Leveraging Artificial Intelligence in the Laboratory" from Roche Lab Leaders, September 23, 2023. Downloaded January 5, 2025 from: https://lableaders.roche.com/global/en/articles/leveraging-artificial-intelligence-laboratory.html.

3. Finally, the implementation of AI-driven technologies in laboratory settings introduces new vulnerabilities, as sensitive data becomes a valuable target for cyberattacks.

It will be key to strike a balance between harnessing the benefits of AI whilst ensuring data and patients are well protected (Lab Leaders, 2023.)

## How generative AI helps to improve quality[5]

At Bosch, during the manufacturing process at the Hildesheim plant, photos are taken by an integrated camera, not of the associates, but of the stator, which is an electric motor component made at the plant. Reimer, the project Lead states that it's all about images. Not all of them are real. When the Hildesheim plant decided to start up a production line for a new product type, it asked the question: How can we get enough images of all defect types — without actually producing defective parts? The answer was generative artificial intelligence (AI). (Bosch, 2023.)

"Thanks to generative AI, we can now train our models for automated optical inspection at a much earlier stage, which makes our quality even better," Riemer says. The plant expects that project duration will be six months shorter with the new approach than with conventional methods, leading to annual productivity increases in the six-figure euro range. The first pilot line was launched in Hildesheim at the end of 2023, and another one followed at the beginning of this year. "There are sometimes hundreds of welds per stator," Schwarz, an engineer on Riemer's project team and an expert in image processing, says. "While humans catch errors at an average rate of 70 to 90 percent, a finished AI model achieves almost 100 percent." Has he ever mistaken a real picture for an artificially generated one? "Of course," he says, grinning. "And vice versa, too."

5 "Generative AI In Manufacturing: Out of The Old, Emerges The New," downloaded on January 6, 2025 from: https://www.bosch.com/stories/ai-image-recognition-production/.

### Amazon Uses Automation to Assist Employees and Deliver For Customers[6]

Amazon is using two robotic systems, Sequoia and Digit, to help employees respond in a better and faster ways to customers' orders and support workplace safety. "We now have over 750,000 robots working collaboratively with our employees, taking on highly repetitive tasks and freeing employees up to better deliver for our customers." Amazon says! It adds that "with all of these efforts underway, it makes us particularly proud to share that we have just launched a new robotic system in time to help fulfill customer orders for holiday shopping this year. Sequoia is the name of this new technology, and it's now operating at one of our fulfillment centers in Houston, Texas."

What are the benefits of this robotic system and automation in the quality that Amazon provides? The article lists these benefits:

- **Better Customer Satisfaction**. By reimagining how we store and manage inventory at our sites, Sequoia will help us delight customers with greater speed and increased accuracy for delivery estimates, while also improving employee safety at our facilities.
- **Faster Order Fulfillment**. Sequoia allows us to identify and store inventory we receive at our fulfillment centers up to 75% faster than we can today. This means we can list items for sale on Amazon.com more quickly, benefiting both sellers and customers.
- **Shipping Efficiency**. When orders are placed, Sequoia also reduces the time it takes to process an order through a fulfillment center by up to 25%, which improves our shipping predictability and increases the number of goods we can offer for Same-Day or Next-Day shipping.

6 "Amazon Announces 2 New Ways It's Using Robots to Assist Employees and Deliver for Customers." Downloaded on January 5, 2025 from: https://www.aboutamazon.com/news/operations/amazon-introduces-new-robotics-solutions.

- **Storage Optimization**. Working as a set of integrated robotics systems, Sequoia will dramatically simplify and optimize how we store goods and help employees pick those goods in a safe way.
- **Reducing Incident Rates.** This collaboration is driving meaningful impact: Company data shows that, in 2022, recordable incident rates and lost-time incident rates were 15% and 18% lower, respectively, at Amazon Robotics sites than non-robotics sites. Sequoia will help continue this positive trend.
- **Mobile Solutions.** Amazon Robotics continues to invest in and try novel approaches to automation. That's why we also have begun testing mobile manipulator solutions at our robotics research and development site just south of Seattle. These are mobile robots that can move while also grasping and handling items.
- **Better Use of Technology.** Our initial use for this technology will be to help employees with tote recycling, a highly repetitive process of picking up and moving empty totes once inventory has been completely picked out of them.

## Success Stories from Implementing ERP Systems

In her "3 Successful ERP Implementation Case Studies," Suzy Strutner presented three good case studies of ERP implementations. She started by defining ERP (Enterprise Resource Planning) implementation as "the process of installing ERP software, and then making sure your team is making the most of the technology." It is relevant to this topic to understand that this ERP process consists of three phases. Strutner (2023) listed these phase as 1) Installing onsite software and/or contracting with a SaaS provider. Often, companies seek help from the vendor or an implementation partner specializing in ERP, the vertical industry or specialized use cases, 2) Migrating data from your previous system or systems into the new ERP, and 3) Training the finance team to use the new software. That process may include knowledge transfer from employees who previously used the system or formal education from your implementation partner.

The following are descriptions of these three cases as presented by the author.

### Case Study 1: Fulton & Roark[7]

Fulton & Roark, a retailer of men's grooming products, is an example of a successful ERP implementation. Prior to upgrading to full-featured ERP, the North Carolina-based business tracked its inventory in a spreadsheet and its financial data in desktop accounting software, Sage Live. When the company began doubling sales year-over-year, leadership felt its current processes weren't keeping up. Spreadsheets couldn't account for changing inventory costs, and the accounting software didn't have the workflows necessary to record the cost of goods sold (COGS), an important financial metric. As a result, the Fulton & Roark team did double data entry—manually.

To centralize all work in one place, the company's co-founders implemented NetSuite ERP. After a three-week implementation process, changes were immediate, according to team members. Finally, the Fulton & Roark team was able to 1) Catch and correct bookkeeping mistakes related to inventory, 2) Stop working with external accountants, growing both unit and dollar volumes significantly with no extra headcount, 3) Increase sales roughly 50% year-over-year without increasing headcount, and 4) Get a more accurate picture of margins and inventory, which helped grow its ecommerce operation.

### Case Study 2: N&N Moving Supplies[7]

N&N Moving Supplies, a family-run distributor of moving equipment and supplies, successfully implemented an ERP system after expanding from one location in Georgia to three locations in multiple states and more than quadrupling its workforce.

Running on QuickBooks and a third-party payroll provider, it was nearly impossible to keep accurate time records and reconcile payroll with general ledger accounts. So, N&N turned to ERP to manage its accounting and payroll processes. During the implementation process for NetSuite ERP, N&N worked with a NetSuite partner, NOVAtime, to add a time-clock solution to the main system. With this in place, N&N was able to 1) Reduce payroll processing time by 84%, 2) Balance accounts faster, 3) Improve the accuracy of hours and vacation time counts, 4) See labor-cost trends across its three locations, and 5) Give employees access to personalized ERP dashboards on iPads at each of its sites.

### ERP Case Study 3: Green Rabbit[7]

Once upon a time, two friends launched a candy wholesale business. Then, they ran into a problem: shipping melt-prone chocolate bars during summer months. After the pals designed a heat-sensitive supply chain, other food companies turned to them for help. The duo rebranded as Green Rabbit, a supply chain logistics provider specializing in fast delivery of perishable goods.

That shift in strategy tested the company's current processes, which relied on QuickBooks, Excel and email. Green Rabbit's multiple databases couldn't communicate with one another, making real-time data analysis impossible. The warehouse was often waiting on help from the IT team, too.

Green Rabbit chose NetSuite ERP, and the NetSuite professional services team got the company up and running on the system in three months. Green Rabbit now 1) Helps customers ship candy, snacks and more across the country in 24 hours from one of three different warehouses, without inventory errors, 2) Delivers tens of thousands of orders per day, without delays, 3) Gets guaranteed error-free data from its ERP system, instead of risking errors from manually entered data, and 4) Could triple its order volume, if desired, without impact to its systems.

---

7 Suzy Strutner, "3 Successful ERP Implementation Case Studies," ORACLE Netsuite, January 2021. Downloaded from https://www.netsuite.com/portal/resource/articles/erp/erp-implementation-case-study.shtml on January 7, 2025.

# Chapter 10
## The Future of Quality 4.0

Defining the future of Quality 4.0 in a specific company or industry is an interesting thought exercise because the technologies available vary widely. The methods to use data include many different analyses and approaches that vary across industries. Determining the long-term future of Quality 4.0 starts with a clear look at how past industrial revolutions have changed the landscape of business and an ideal future state can be conceptualized for this industrial revolution.

### *The First Industrial Revolution*

Steam power enabled factory growth and production flexibility, allowing mills to locate anywhere and facilitating faster product distribution via steam trains.

### *The Second Industrial Revolution*

Electricity revolutionized operations, with the lightbulb enabling 24/7 factory work, transforming delivery methods, and optimizing management processes.

### *The Third Industrial Revolution*

Programmable logic controllers (PLCs) introduced automation, improving production consistency and speed without altering product nature, and focusing on delivery methods.

## Industrial Revolutions: Changing How We Operate

Much like the past three industrial revolutions, the fourth industrial revolution is changing how day-to-day operations incorporate data into processes to shift decision-making and actions. There's a shift in thinking about how a more data-integrated future of business operations will likely result in companies that focus heavily on the optimization of standard operations. The ability to monitor performance in real time, or even to predict performance in the future, opens the door to more flexible businesses that are agile in their ability to respond to valuable information. The ability to pivot throughout the day to optimize quality and system performance is the operational norm of the future with Quality 4.0 and other Industry 4.0-based systems. Think about how an optimized future can help set the vision for the long-term use of the Quality 4.0 systems we architect.

Ultimately, those day-to-day information use cases should be tied back to the overall goals and mission of the company and should relate directly to the stated quality goals of the company as well. A company that strives for zero defects should emphasize systems that perform inspection and ensure that non-compliant parts don't advance through the production system. Companies that prioritize things like innovation and service should set their systems up to emphasize customer connection and feedback. Shortening the time between information generated outside of the company; information gathered from customers; and decisions within the company that improve the company's performance and *perception* of will be a popular system deliverable in the future.

How companies operate encompasses many factors. Leadership styles, communication styles, resource allocation, project priorities, strategic goals, and so many other factors roll into the day-to-day operational modality of every aspect of a company. Quality 4.0 systems should be designed with those business factors in mind, with emphasis on the users and use cases of information. The only way to successfully shape a long-term future process that uses new technology is to think about the people and processes supported by that technology.

## Implementing Technology

Most technology implementations in industry focus on the time required for an ROI. Most projects are planned and diligently managed through implementation only; then, they transition to different groups of employees to maintain results or drive continuous improvement. With all the focus on what can be achieved with technologies and how to go about implementing them, it is worthwhile to have a plan for the long-term changes that come with major technology overhauls to a business. A Quality 4.0 system will change how decisions are made and improve day-to-day information management, if done correctly. This long-term benefit should be encompassed in the initial vision of a Quality 4.0 project so the system can support the desired sustained improvement and success.

Consider the history of quality. Throughout history, technology has driven changes to how things are done, but that technology doesn't necessarily change our goals or what we would like to achieve. Agriculture has been central to the history of civilization, and we still farm. Farming is now done on an industrial scale and requires a smaller workforce to feed the world, but it is still a critical industry to our overall success. The "4.0" in Quality 4.0, of course, describes the fourth industrial revolution. Thinking about the changes that occurred during the first three industrial revolutions should be seen as a blueprint for the long-term vision of the fourth. The widespread use of steam power, electricity, and programmable logic changed everything about how work is done, and data technologies will have the same long-term impact. By the end of our careers, how our companies' goals are achieved will be very different.

## Defining the Future State

Knowing that the fourth industrial revolution changes how we use information by integrating data collection into the process of running the business, a future state of Quality 4.0 can be defined. Two elements must be considered for any business:

1. The core competencies of the business and the quality that is offered to customers
2. The way decisions are made around quality and the way information is used

The ideal future state for Quality 4.0 uses technology to integrate these decision processes into an automated system that improves the quality performance of a business in the context of the value that the business delivers to customers.

While the value proposition of businesses differs, the concept of delivering value and making quality a core deliverable is nothing new. Specifically, lean manufacturing focuses on the value stream and the value proposition, and TQM makes quality one of the core deliverables of every position. Architecting a Quality 4.0 future state should leverage existing cultures and methodologies. Deploying new technologies is challenging for a variety of reasons, and reducing overall complexity by architecting tried-and-true methodologies into the approach is worthwhile. Leveraging what is known about existing methods and cultures should be a part of designing the information flows that will make a Quality 4.0 system worthwhile. Each of these elements can be defined individually but should be synthesized into a final system.

## Core Competencies: The Value Proposition

Implementation of Quality 4.0 systems will likely begin with singular applications: a vision system to perform 100-percent inspection; an automatic alarm system that communicates maintenance needs; a dynamic production schedule that updates with real-time customer orders; or any number of other performance improvement applications of the data > analysis > information > decision > action model of data strategy development. These initial applications must center on the value proposition of a process, as already detailed. The broader approach to the implementation of Quality 4.0 is the full-scale digital transformation of a business. By following the same model of data strategy development, the central pillar of a Quality 4.0 system becomes the core competency, or competencies, of the business being improved.

Understanding and defining a business's core competency requires the same style of thought and consideration as identifying the value proposition of a process. Start by asking what unique product, process, service, or other factor is driving the success of this business on the market. The core competency may be that the business is a cost leader, it may be the service level and professional reputation, or it may be the unique product offerings.

There will be no one-size-fits-all answer to what a business's core competency is, and, in all likelihood, no two businesses will survive based on the same identity. Take, for example, a hospital as a business. Geography is one of the core competencies of a hospital. Many of the users of a hospital, the customers (who are more appropriately called patients), arrive through the emergency department. In most emergencies, people needing healthcare select an emergency medical service provider based on where they are and how fast they can get there.

In other cases, brand loyalty can be a core competency. Take pickup trucks as an example; customers who drive them often have extreme brand loyalty. Sometimes brand loyalty with trucks is generational, with an individual driving a specific brand because their parent or even grandparent swears it is the best truck on the market. The key to identifying the core competency to be supported with a Quality 4.0 application is to gain an understanding of the business at hand and the customers who keep it profitable.

Once a set of core competencies and values has been clearly identified within a business, the Quality 4.0 system vision can be developed fully. Ideally, the vision is conceptualized and developed with company leadership before Quality 4.0 implementation projects begin, but this may not be the case if a project manager starts with the scope within their influence. Regardless, the Quality 4.0 vision should start, like a data strategy at any level, with understanding the actions that deliver value, the decisions that drive those actions, the information that feeds those decisions, the analysis that informs those decisions, and ultimately, the business level data that supplies that analysis. This method of Quality 4.0 development and deployment functions at all levels. The implementation model should be scaled to encompass the entire business if the goal is to undergo a full-blown digital transformation.

Transforming an entire business is a big undertaking, and it goes well beyond just technology. With the barriers and implementation methods identified within this book, it should come as no surprise that at the enterprise level of a business, business culture is significant to Quality 4.0 success. Changing culture is a nuanced and often challenging task, which is covered in other books. One of the most important aspects of getting the culture in the right place is identifying what that culture needs to be. It is nearly impossible to decide how to install or change a culture without a target end state in mind. Just like any complex task, it is important to map the finished state before attempting to take the first several steps.

## Quality Cultures and Quality 4.0

The identification of the right culture to maximize value from a Quality 4.0 system is a critical strategy step at the outset of system implementation, and it is central to the strategy and implementation of the entire system. Culture and the approaches to changing culture are the main components of the strategy and implementation of the second pillar in the deployment model. This side of the model is what enables value to be realized and implementation to be successful (Figure 10.1).

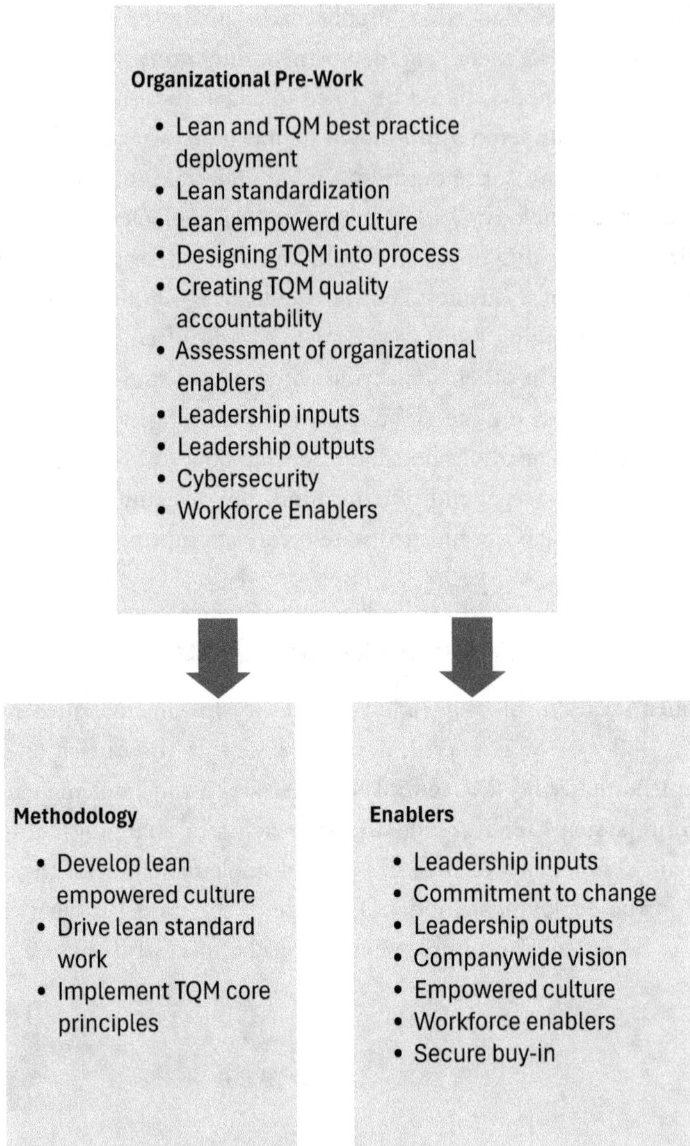


**Figure 10.1** Conceptual model for implementation.
Source: Adapted from Jones (2023).

Many work cultures exist, from varying leadership and management styles to communication styles, matrixed organizations, traditional organizational structures, agile methods, and other styles that could be explored and detailed. To understand which cultures may work best with Quality 4.0 systems, it is important to explore just how a Quality 4.0 system is going to enhance the operation of a business, and to work with the cultural elements that suit those strengths and improvements. The financial benefit of a Quality 4.0 system, as detailed in Chapter 8, serves as a baseline to define how an operation is advanced with Quality 4.0. In all examples, the value of Quality 4.0 comes from the improvement of information (either better, faster, or more) that comes from the system. As described in Chapter 4, in detailing a data strategy, information makes a business or process perform better by informing decisions and driving actions. These decisions and actions must be better than the baseline operation of the business or process to be worth pursuing, which means the decisions must be better (resulting in a higher success rate of the actions) or the decisions have to come faster (resulting in less lag time before realizing the benefit of the actions).

Better decisions come from better information and drive better actions. "Better" may mean a reduction in the cost of poor quality by reducing defects produced by a product. It may refer to improving the perception of the service level of the business in the eyes of customers. It may result in more consistency in the process, which gives better forecasting and planning. There are many ways that businesses can be better, and maximizing the opportunities to perform better, collaborative efforts, and companywide accountability are generally good cultural elements to install.

Faster decisions are much more straightforward than better decisions. To define "faster" is simple: a system is currently producing information that drives decisions and actions, and the new system should provide the same information, driving the same decisions and actions, faster under the same conditions. This makes logical sense and can easily be related to the automation of a Quality 4.0 system to create information in real time or near real time. The cultural element, however, is that when a decision is made (and decisions can be automated; the system can present information and use that information to choose an action in some cases), an action still must occur. This is most notable in non-conforming conditions, when actions must be made to contain and correct issues. The action must be taken by a person, and to realize value from the information and decision being accelerated, the user of the information must be available to act faster as well.

The easiest way to allow for actions to happen faster is to install trust in the system and empower people to use information and take action. Creating a workforce in which employees are empowered is captured in Figure 10.1 with the direction to install a lean empowered culture. This is further supplemented with the quality accountability that is assigned to the entire workforce through TQM.

Collaboration, accountability, and empowerment: These three cultural ideals are central to maximizing the value realized from a Quality 4.0 system. Defining and changing culture is challenging. When the opportunity arises to utilize best practices that are proven and have stood the test of time, it is worth doing so. In this case, the elements of collaboration, accountability, and empowerment relate very well to two methodologies that have been well-defined and explored for years: total quality and lean.

## Total Quality

Total quality is a quality philosophy that encompasses the entirety of the organization. The core idea with Total quality is that quality is not an independent function that inspects parts or services customer complaints, but a part of everybody's job that must be architected into the entire business and its processes. Total quality covers all three aspects of the ideal Quality 4.0 culture. Collaboration is necessary when the priority of the system is to achieve good quality and that priority is built into the function of everyone and every task. Everyone must collaborate on quality goals as a key strategic output of the company. Accountability is also very clearly built into the process in a total quality environment, as everyone is held directly accountable for the quality output of their process and tasks. Empowerment must come with accountability, or the "accountability" falls apart and would ultimately be nonexistent. If individuals aren't empowered to act under adverse conditions, then being accountable to the outcomes of those conditions is inappropriate and can't succeed. Total quality correlates with success in digital transformation because the Quality 4.0 system builds in observation and self-monitoring. These two ideals work very well together, as the logical extension of creating a system that observes and predicts its own performance is to empower employees to respond to the information about that performance and drive improvement. Empowerment builds quality into the role of everyone and gives them information to take the best possible actions and trust the system instead of needing to wait for the direction of management (Babatunde, 2020).

## LEAN

Many people think of lean in terms of inventory management, kanban systems, and single-piece flow, but this is only the technical side of lean. Lean manufacturing involves a set of tools and methods designed to achieve a lower WIP, and a more consistent and reliable production system. The ultimate goal of single-piece flow is to achieve a production system that produces only one piece at a time with no ballooning inventory between operations—no capital tied up in parts that are paid for and sitting on the shop floor.

Beyond the technical side of lean is the lean culture, which is actually critical to the success of the lean toolbox. lean culture centers on concepts such as kaizen (rapid change for the better), Gemba (the real place), and Andon (stop the line). A kaizen culture is a continuous improvement culture with the goal of involving the entire workforce to make small, incremental changes frequently. These frequent improvements drive a high-performance system over time; it is an incredibly effective method of improving a process or system. Gemba, the real place, is the concept that everyone needs to invest in supporting the value-added work that the customer desires and is willing to pay for (which is the lean definition of value). To support value delivery, leadership and other roles who routinely do work at a computer—sometimes far away from the shop floor where the value is delivered—must visit the Gemba frequently to supply support for problem-solving and improvement efforts. Finally, the Andon principle is that everyone can stop the line. This principle is incredibly important, as a single-piece flow system will fall apart if a defective part is sent down the line. The purpose of Andon is for the line to stop when a deviation is identified so that the team can address the cause of the defect as fast as possible and maintain a high quality level.

These principles overlap with the collaboration, accountability, and empowerment model of Quality 4.0 culture very well. Kaizen ideology demands collaboration to be successful, as improvements can and should come from all individuals throughout the organization. Kaizen culture also cannot exist without empowerment, as the only way to reliably get participation and ideas from a workforce is to empower them to take ownership of those ideas, drive them forward, encourage pride in the ideas, and provide recognition and reward appropriately. The fastest way to shut down a broad, continuous improvement culture is to shut down ideas that come from front-line employees.

The culture of Gemba embodies all three principles. Collaboration is driven by the fact that leaders must visit the floor to support solving problems. Leaders are at the Gemba to provide support and resources for those employees who are directly adding value. Gemba also drives accountability by putting the real work on display and providing transparency. People cannot hide away in a facility that embraces Gemba, accountability is built into the process. Finally, empowerment comes from those leaders supporting the problem-solving culture as problems arise. Leaders should support problem solving in a Gemba culture by providing needed resources; however, the goal is not for leaders to categorically solve every problem or dictate the solution; often the simplest solution is to empower the localized employees to act on the ideas they already have when a problem is identified.

The Andon approach also pushes all three elements of the Quality 4.0 culture. Allowing employees to stop the line is the ultimate form of empowerment; stopping the line stops production. When the line is stopped, it is the responsibility of all support functions to collaborate in solving whatever problem caused the line stoppage. Employees empowered to stop the line are held accountable to not send defects downstream to future processes, which would just cause them to waste capacity on processing defective parts. These cultural aspects of lean are worthwhile to implement, whether a digital transformation is on the horizon or not—and they are critical as groundwork to support a future digital transformation.

## Realizing Long-Term Value from Quality 4.0

Consider a Quality 4.0 system in which the business is fully mature with Quality 4.0. That might involve a shop floor where every in-process specification is monitored automatically. Any time a part has a defect, people are alerted. Those people are empowered to respond immediately. The line stops when defects are found, and the right individuals are sent to respond to the issues. The labor that was previously assigned to collecting quality data and doing trending and analysis is now reassigned. Those individuals are recipients of completed analysis and actionable information. That analysis compiles to assess the overall performance of the system, which provides information on which issues deserve resources and prioritization to be solved. The former quality control employees are now focused on continuous improvement, taking on small projects and completing Plan, Do, Check, Act (PDCA) work to improve those issues identified by the system.

This is what a high-performance Quality 4.0 system would look like on the shop floor. These concepts can also be applied to the support functions of the business as well. The supply chain can be optimized, and the planning and forecasting functions can be better integrated into the system. Even outside of traditional manufacturing, the system can be observed and information can be automated. All these objectively measured improvement opportunities and improvement projects will begin to add up over time to drive the performance of the business to the next level. This is where the true value of a Quality 4.0 system will live. The team that executes the most small improvements with the labor that is freed up from automation of the information flows is the team that will be the most financially successful in the next generation of business.

## A Thought Example: The Perfect Cup of Coffee

It often helps to think through the tangible aspects of how a concept will work with an illustrated example. For this example, the goal is to deliver the highest quality cup of coffee at a café. The big-picture process steps are:

1. A farmer grows the coffee beans.
2. Beans are harvested.
3. Beans are shipped.
4. Beans are roasted and ground.
5. Grounds are prepared and shipped.
6. A café makes a coffee recipe and presents to the customer.

In a traditional coffee supply chain, these steps have many gaps between them. Typically the café would handle its own quality function with the customer-facing aspects in mind. The manufacturer who handles roasting and grinding would act as an independent function in this process, with the café as their customer. The company doing logistics would be a separate entity, and none of the organizations involved in later steps would directly correlate back to the farmer who handles growing and harvesting. Each of these systems has its own quality metrics, goals, and systems.

The farmer most likely considers yields, rain, costs of fertilizer, etc., and they are very much at the mercy of the weather when it comes to the quality of the finished product. It is also worth considering that the variance in grown coffee beans will be significantly greater than the variance in an engineered product coming out of a factory. Beans could vary in taste, size, ripeness, and many other factors. Quality, from the perspective of the farm, most likely comes down to the beans being harvested with minimal external impurities and good yields from the acreage dedicated to growing the beans.

The shipping company is obliged to protect the beans in transit. They care about efficiency, economies of scale, and meeting deliverables. They likely have more mature quality metrics than the farmer, but there is still variance in the logistical system. If the beans are shipping from Brazil, a major coffee exporter, to the U.S., then there is an element of dependency on the weather conditions for them as well. The most important thing for the shipping company's quality is that the beans are not damaged or otherwise rendered useless in transit. The shipping company and their quality contribution to the cup of coffee is the same for the raw beans and the ground, processed, beans.

The roasting and grinding operations are full-blown factories. They operate under the scrutiny of groups like the FDA, with very mature quality measures and metrics. They likely have traceable records of both product and process quality, although much of the data collection and analysis is likely manual at this point. They have the most control over their shipped product with the most repeatability prior to delivery to the shipper.

The café is likely the smallest business of them all (unless it is a chain). It has a lot of flexibility in the quality of delivery, as elements such as service and atmosphere probably matter to the customer. The recipe for how a specific coffee drink is made is likely consistent, but customers tastes for food and drink can shift and vary widely, as can what a customer considers "ideal" quality for a cup of coffee at a café.

In this theoretical production system that encompasses the full life cycle of a cup of coffee (hopefully the perfect cup), where is quality determined? Is it different for each step in the process? Is it all dependent on the experience of the individual who drinks the coffee? How can it be improved when so much depends on uncontrollable factors? Consider a specific example of extra rainfall prior to harvest. This may cause the beans to be less ripened by the time they are harvested. This may cause the grounds to retain slightly more acid, which may change the flavor of the coffee and the customer experience. If the lead time on this supply chain is several months, that could result in a long time to wait to see if the product was good and usable.

Now the bigger question is: What can Quality 4.0 do? In addition to being able to deploy quality systems as described in the previous chapters of this book at each step within the supply chain, there is a unique opportunity to create a global forecasting model within the entire supply chain by linking data together. This is only plausible due to the automated and connected nature of Quality 4.0 systems. Doing this level of data management manually would likely be untenable, but with a comprehensive Quality 4.0 system process, performance data from every step of the supply chain may be linked together to begin to build a comprehensive and predictive model for what quality of coffee will be served at the café.

First, the data from the farm itself (weather, fertilizer use, what strain of bean was planted, the general nutritional value of the soil geographically where the bean was farmed, and so many other factors) can be built into a baseline data traceability package that goes with the beans. The same can be done for the logistics company (was the bean sent over land or by sea? Does the sea air affect the flavor profile due to the salt? And so on). The manufacturer supplies information on batch sizes, recipe parameters for roasting and grinding, etc. Finally, the café is where all that data is linked to the customer. A feedback model can be generated by plugging into social media and harvesting reviews made by these customers.

In a future state, the café may optimize sourcing for the customer base it has, offering recommendations for customers due to the understanding of how different process conditions impact the flavor and reception of the cup of coffee. Aspects of the delivery will begin to improve, and the overall quality can increase significantly. This example highlights the goal of centering on value and core competencies, as it is valuable to link quality performance with the ultimate customer.

This future state idea/thought exercise is intentionally vague; it is here to generate thought about what may be possible. This big picture approach to data, how it should be used, what information may be valuable, and why it's worthwhile to think outside of the box is incredibly worth it. When conceptualizing a digital transformation, value proposition, core competency, or data strategy, asking and exploring what all can truly be done today can really help maximize the value realization from the overall system. Always ask what information is worth having, even in cases when it is not completely obvious how that information may be generated.

# Chapter 11
# Transitioning to Quality 5.0

As Quality 4.0 is becoming clearer to practitioners, questions are arising about Quality 5.0. The next round of advancements is already being conceptualized, and the foundations lie within the Quality 4.0 space. Much of the current Quality 4.0 vision draws roots from the third industrial revolution (PLCs), when factories started operating with greater degrees of automation. Similarly, the relationship between 4.0 and 5.0 is changing as technology advances at lightning speed.

As it stands, Quality 5.0 still requires definition and clarity as a term. The same logic that defined Quality 4.0 applies. Quality 5.0 represents the marriage of Quality with Industry 5.0. The European Commission (Directorate General for Research and Innovation, 2022) defines Industry 5.0 as taking the competencies of Industry 4.0 technologies and applying them to create businesses that are:

- Sustainable
- Human-centric
- Green
- Resilient

The beauty of this definition is that it is a natural extension of the Quality 4.0 deployment methodology outlined in this book.

## Feeding into Quality 5.0

Implementing Quality 4.0 as described in this book flows directly into a Quality 5.0 future. It is important to understand the long-term destination of the businesses that will undergo these transformations, as more and more data becomes usable to greater degrees; companies will connect on much more personal levels with customers. The future of data-driven business is a future in which customers' individual needs are more visible to the businesses they choose to patronize. The days of Henry Ford's Model T, which could famously be purchased in "any color as long as it's black," are gone. The original assembly line leveraged mass production to satisfy a market demand for "vehicles," which was incredibly successful at the time and planted many of the seeds of today's manufacturing best practices. The 4.0 auto manufacturer of the future will provide the same vehicles to the same customers, except each customer's individual needs can be configured and met. Customizable, flexible manufacturing is the clear future of Quality 4.0.

This flexibility to make exactly and only what customers want is inherently more sustainable, human-centric, and green all at once. Overproduction for economies of scale has been a major contributing factor to the unsustainable and environmentally damaging practices of the past. This type of manufacturing is also design-centric, as the focus is on the design of products that will satisfy most customers and the design of manufacturing systems that can be profitable. Connecting all the way to the end customer puts that customer at the very center of all of the decisions for design, production, cost, quality, etc.

Another view of human-centric action comes from the implementation model proposed in this book. Quality 4.0 will not be successful if it's implemented with a software-first approach by IT professionals. Success requires employee consideration, input from a cross-functional team, and most importantly, employee empowerment. If employees are not empowered to use the information provided to them, the system will not maximize value from data (see Chapter 8). A successful Quality 4.0 implementation drives involvement and empowerment; it naturally creates a more human-centric manufacturing system in which the technologies must enhance an employee's ability to succeed.

Resilience comes as a byproduct of rolling out Quality 4.0 as opposed to a different Industry 4.0 strategy. As the manufacturing system becomes more and more customized to customers' specific needs and wants, quality also must be a core deliverable. Many business systems have succeeded throughout modern history, and largely they can be categorized as:

- Quality-centered
- Service-centered
- Cost-centered

Quality and service often go hand-in-hand for businesses, as it can be challenging to prioritize one without the other and be successful. Quality 4.0 should lead to customer-focused service as a priority for the business. As customers have a more tailored experience, it cannot come at the expense of quality and service. Industry 4.0 approaches such as Logistics/Supply Chain 4.0 will likely become cost leading businesses and will likely be very successful strategies for the right cases—but the businesses that can provide customized needs for each customer while maintaining high quality and service levels will be resilient against many external forces. Quality inherently drives resilience into a business.

## Implementing Quality 5.0

It is premature to provide an implementation model to Quality 5.0; however, beginning with Quality 4.0 will set many companies up for Quality 5.0 success. There is one other aspect to consider with Quality 5.0 that is not directly addressed in our current understanding of Quality 4.0: ethical implementation of technology. It is a key factor to Quality 5.0. As business processes become more and more autonomous, the labor force will adapt in skills and the nature of jobs will evolve. To be sustainable and human-centric, the outcomes for those humans must be considered in the design and strategy phase of a Quality 4.0 deployment to become truly Quality 5.0. An entire book could be authored on the ethics of AI and automation (and likely it has), but a high-level set of principles to guide a practitioner from Quality 4.0 to Quality 5.0 is as follows:

- Design systems that help employees succeed, not just displace them through automation.
- Ensure that employees are set up for a future with automation, including training.
- Consider the impacts of technology on both upstream suppliers and downstream customers to design systems that benefit everyone.
- Consider the environmental impact in gaining efficiencies, and design future systems to waste less while being more customizable.
- In general, if technology provides a massive benefit, design systems where that benefit is distributed across stakeholders such as customers, suppliers, and employees in equitable ways.

## Quality for Humanity

In its report about Industry 5.0, the Directorate General for Research and Innovation of the European Commission (EC) (2022) describes the main building blocks of Industry 5.0. The report's findings are as follows:

- Industry 5.0 contributes to the technological update needed for industry to affirm its purpose as a solution provider for our society and as an attractive employer for young talents in search of a purposeful professional life.
- Digitalization provides industry with unprecedented opportunities. Digital technologies such as artificial intelligence (AI) or robotics allow radical workplace innovation, optimizing human-machine interactions will capitalizing on the added value human workers bring to the factory floor.
- By developing innovative technologies in a human-centric way, Industry 5.0 can support and empower, rather than

replace, workers; increase industries' resilience and make them more sustainable.

- Common environmental goals can only be achieved by incorporating new technologies and rethinking the production processes to account for environmental impacts.
- A transformed industry will also have a transformative impact on society. This is especially true for industry workers, who may see their role changed and/or require new skills. The transition to Industry 5.0 will require action in a whole range of areas.

Quality 5.0 is a product of Industry 5.0 that builds upon the technologies of Quality 4.0 and, at the same time, adds new dimensions focused on human-centricity, sustainability, resilience, and social responsibility. (See Figure 11.1.) Quality 5.0 aims to integrate the latest technological advancements of Quality 4.0 with a renewed emphasis on human values, sustainability, and societal impact. Quality 5.0 is characterized by five components that evolve around:

- Human value
- Sustainable environment
- Resilient social impact

Several studies have provided guidance and insights on the characteristics, benefits, success factors, and challenges of Quality 5.0 as discussed below [Arsovski (2023), Frick and Grudowski (2023), and Farooq (2024)].

**Human-Centric Approach.** This approach is the collaboration between humans and AI, including robots and other automated machines or equipment. While AI handles data-driven decision-making and routine tasks, humans focus on creativity, critical thinking, and value-added activities. As a core value of the European Commission (EC), organizations prioritize employee well-being, engagement, and continuous learning, recognizing that motivated and skilled workers are essential for achieving high-quality outcomes.

**Sustainability and Social Responsibility.** Quality 5.0 incorporates sustainability into quality management practices, ensuring that products and processes minimize environmental impact and promote ecological balance. In addition, a strong emphasis on ethical business practices is essential, along with transparency and corporate social responsibility for society.

**Resilience and Adaptability.** Resilience and adaptability are two components that, together, create an agile organization that quickly adapt to changes in the market, technology, and regulatory environments to remain competitive.

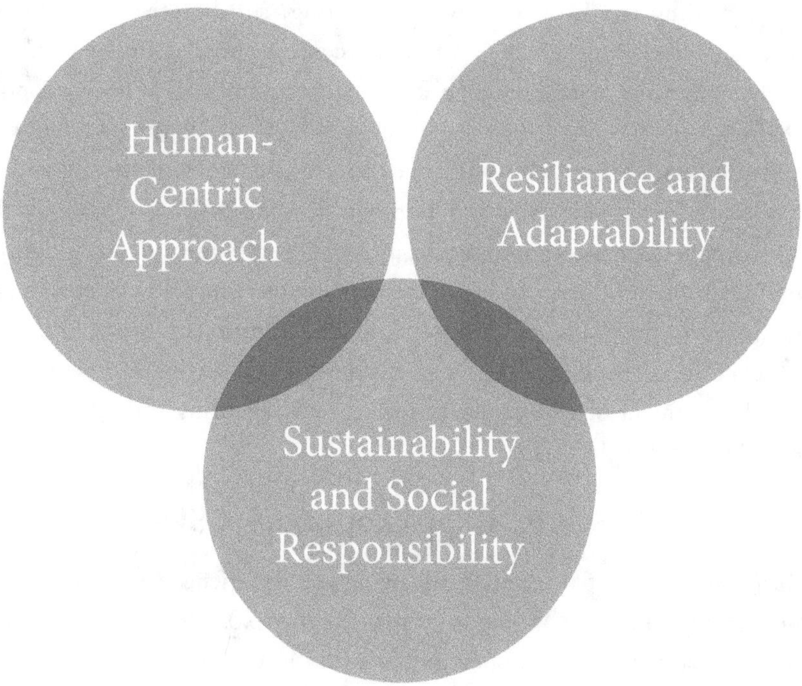


**Figure 11.1** Characteristics of Quality 5.0.
Source: Prepared by author based on material from Monika Hattinger and Kostas Stylidis (2023). "Transforming Quality 4.0 towards Resilient Operator 5.0 needs." 56th CIRP Conference on Manufacturing Systems, CIRP CMS '23, South Africa. *ScienceDirect*: *Procedia CIRP* 120 (2023) 1600–1605.

Quality 5.0 represents a significant shift in quality management, blending advanced technologies with a human-centric and sustainable approach. Successful implementation requires overcoming cultural resistance, integrating new technologies, bridging skill gaps, and maintaining a focus on sustainability and ethical practices. By fostering strong leadership, engaging employees, and leveraging the right tools and technologies, organizations can achieve superior quality outcomes that benefit both their customers and society at large.

## Quality 5.0 Technologies

Quality 5.0 uses different technologies that are also used with Industry 5.0, including the IoT, 5G, cloud computing, collaborative robots, big data analytics, blockchain, and other connectivity technologies that enable real-time data exchange and analysis across the entire supply chain, thereby enhancing responsiveness and flexibility. Predictive and prescriptive analytics, ML, and AI are used to anticipate quality issues, optimize processes, and improve decision-making. Quality 5.0 also supports mass customization, allowing companies to offer highly personalized products and services that meet individual customer preferences and needs. Because of its human-centric capability, Quality 5.0 continuous feedback loops and customer engagement are integral to the development of new products and services, ensuring they align with evolving market demands.

## Benefits, Challenges, and CSFs

Although Quality 5.0 is not at full swing yet, experience from the implementation of Quality 4.0 indicates that Quality 5.0, with its human-centric approach, has more opportunities to do more for the organizations [see for example, Arsovski (2019), Arsovski (2023), and Saraiva (2023)].

**Benefits.** Focusing on its components of human-centric approach, sustainability, resilience, and social responsibility, Quality 5.0 contributes to improving the overall well-being of employees, customers, and society. Organizations that embrace Quality 5.0 can differentiate themselves through superior quality, innovation, and responsiveness, gaining a competitive edge in the market. In addition, integrating sustainability into quality management helps organizations achieve long-term growth while minimizing their environmental footprint and fostering social good.

**Challenges.** Quality 5.0 represents the next frontier in quality management, blending advanced technology with a human-centric and sustainable approach. By focusing on collaboration between humans and AI, ethical practices, and environmental stewardship, organizations can achieve superior quality outcomes while contributing positively to society and the planet. Challenges that companies need to embrace in Quality 5.0 are the same as those encountered when implementing Quality 4.0 (see Chapter 2 and Adel [2022]):

- Lack of the resources needed to adapt technologies like AI, IoT, and advanced analytics to enhance quality management processes. Integrating advanced technologies like AI, IoT, and big data analytics into existing systems can be complex and costly.
- Lack of a change management process, which sometimes takes a long time to create. To foster a human-centric culture, organizations must prioritize employee well-being, engagement, and continuous learning to harness the full potential of their workforce. Shifting to a human-centric, technology-driven approach requires significant cultural change, which can be met with resistance from employees and management.
- Lack of alignment of Quality 5.0 values with the organizational strategic goals. It is often difficult to balance technological advancements with ethical considerations, and ensuring AI-driven decisions align with organizational values can be difficult.
- Lack of financial capability to embrace the transformation. There is a need to implement agile methodologies

and robust risk management strategies to stay adaptable and resilient in the face of change.

- Lack of proper skills to understand the values of integrating sustainability into every aspect of your operations, from product design to supply chain management. There is often a gap between the current skill sets of employees and the skills required for effective implementation of Quality 5.0. Ensuring data accuracy, security, and privacy while handling large volumes of data can be challenging. Other skills that may be needed and require extensive training and resources are: understanding of digital tools and technologies such as AI, IoT, and big data analytics; proficiency in data analysis and interpretation to derive actionable insights from large data sets; strong problem solving skills to identify and address quality issues effectively; and an understanding of sustainable practices and principles to integrate them into quality management processes.

**Critical Success Factors for Implementation.** These include the following:

- Strong commitment from leadership to drive cultural change and invest in the necessary technologies and training;
- Participative management;
- Involving employees in the transition process, providing them with the necessary training, and fostering a culture of continuous learning;
- Developing a clear strategy for integrating advanced technologies and ensuring interoperability between new and existing systems;

- Continuously engaging with customers to understand their needs and preferences, and integrating their feedback into product and service development;
- Embedding sustainability into the core business strategy and operations to ensure long-term viability and positive societal impact;
- Implementing strong data governance frameworks to ensure data quality, security, and compliance with regulations; and
- Adopting agile methodologies to enable rapid adaptation to changing market conditions and technological advancements.

## More on Quality 4.0 and Transitioning to Quality 5.0

Pedro Saraiva (2023) in his article, "Quality 5.0 in Turbulent Times: How to think about and apply quality in today's world of societal challenges" (*Quality Progress*, October 2023, Volume 56 Issue 10, pp. 18-25), presented an interesting statement about the perception of Quality 4.0 in an organization. Saraiva stated that "although slight differences can be found across authors, in general, this Quality 4.0 wave—driven mostly by and related to Industry 4.0—can be seen as an attempt by quality-supporting organizations that are moving into Industry 4.0 to cover two sides of the same coin (Figure 11.2), corresponding to the exploitation of digital transformation to improve quality management, and quality management to improve digital transformation."

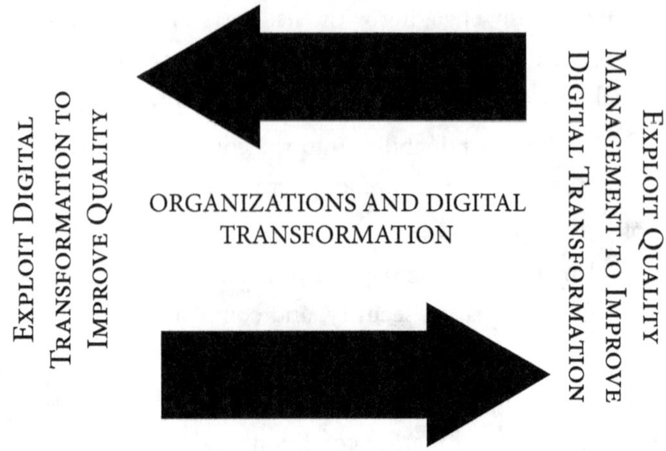


**Figure 11.2**. Organizations and digital transformation.
Source: Drawn by authors based on information from Pedro Saraiva (2023) in his article "Quality 5.0 in Turbulent Times: How to Think About and Apply Quality in Today's World of Societal Challenges." *Quality Progress*, 56 (10): 18-25.

Arsovski (2019) and the Swedish Institute for Quality addressed the need for quality management to better accommodate more of the human-centered approaches and improved quality of life that are major elements for Society 5.0.

## Society 5.0: The Road to a Smarter Society

Society 5.0 is the Japanese version of a better look at the quality of life. Digital transformation—Industry 4.0 and Quality 4.0, for example—has its impact on product development, production, and collaboration. Governments are eager to support this development toward what is often called "smart industries." The Japanese government has taken this one step further, suggesting a vision for a supersmart society called "Society 5.0." Earlier societies, following the four industrial revolutions, included the hunter and gather society, the agricultural society, the industrial society, and the current information society (Bergman, 2022).

The goal of Society 5.0 is to realize a society where people enjoy life to the fullest. Economic growth and technological development exist for that purpose, and not for the prosperity of a select few. Following this concept proclaimed by the government, various activities have begun in Japanese academic circles and in industry. And, although Society 5.0 originates in Japan, its purpose is not just for the prosperity of one country. The frameworks and technology developed here will no doubt contribute to resolving societal challenges worldwide (Fukuyama, 2018). Figure 11.4 identifies the chronological changes in the societal and cultural behavior until the supersmart society 5.0 is reached.

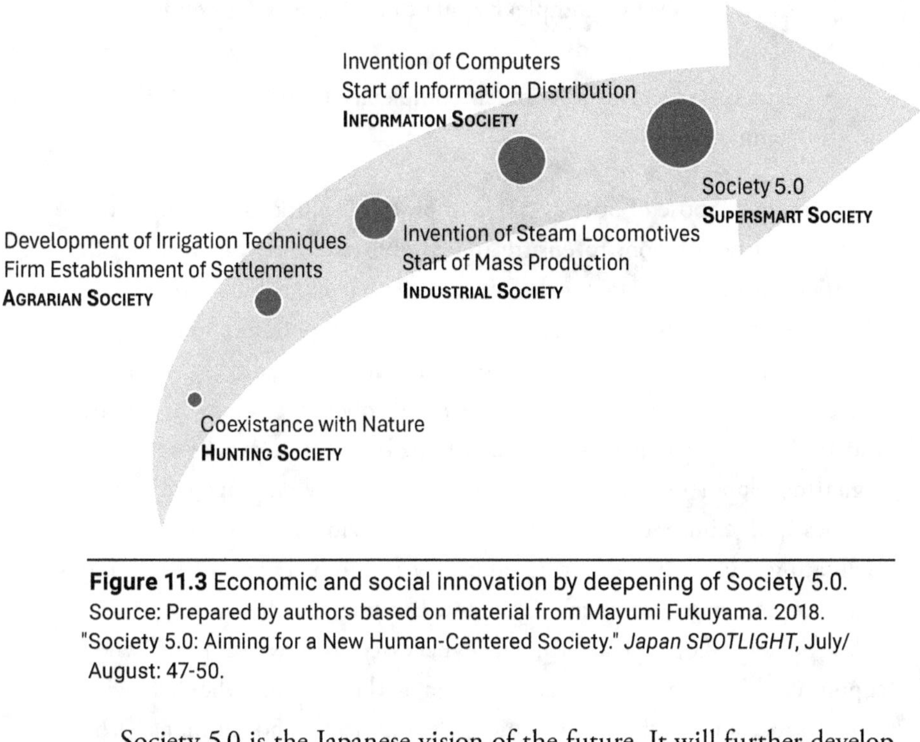


**Figure 11.3** Economic and social innovation by deepening of Society 5.0.
Source: Prepared by authors based on material from Mayumi Fukuyama. 2018. "Society 5.0: Aiming for a New Human-Centered Society." *Japan SPOTLIGHT*, July/August: 47-50.

Society 5.0 is the Japanese vision of the future. It will further develop the information society using all the benefits of digitalization to a "supersmart society "as a solution to challenges in a Japanese context, including decreasing nativity and an aging population (Bergman, 2022).

The purpose of a supersmart society is to create a human-centered society and to balance economic development and prosperity with solutions to social problems by integrating cyberspace with physical reality.

Earlier development for the information society (Society 4.0) gathers information through different networks and is analyzed essentially by human brainpower. Lessons learned are that:

- It is hard for a human to find the relevant information in this excess of data;
- The ability to share information and knowledge is insufficient because of the difficulty human brains have handling the level of complexity due to an increased amount of information; and
- There is a limit to how much complexity humans can handle.

The main purpose of Society 5.0 is to produce solutions to combat the aging world population; to ensure that the virtual and real-world work together; to benefit from the IoT in a way that contributes to society; and to generate ideas for mitigating environmental pollution and natural disasters. In addition, it was observed that with the transfer of business processes to the digital environment, AI and robot technology improved, and the need for manual power decreased. In short, it can be said that digital development and transformation create new employment opportunities and eliminate many business areas. However, AI focuses on facilitating human life, ensuring efficiency in production, and surviving in a competitive environment (Calp and Butuner, 2022).

In Society 5.0, a platform in cyberspace is created that, when supported by AI, can manage big data gathered from the IoT with sensors measuring the environment, equipment, and humans. The potential result is an integration of cyberspace and physical reality, where robots and machines can perform much of the tedious work previously performed by humans.

## Society 5.0 Benefits

The goal is to create a society consisting of overlapping generations with mutual respect—a society where all are given the possibility to live an active and attractive life (Bergman, 2022):

- Humans are liberated from some of our everyday troubles. They have the possibility to be free to do things of their own liking.
- Products and services are created to cover human needs when they arise.
- The social system will be optimized.

There are many possibilities when digitalization is used in all societies. At the same time, we must face the considerable **risks** of such a supersmart society:

- How will a life controlled by AI be perceived when the AI is governed by interests that are not transparent?
- Can AI go its own way and create greater problems for human survival than it solves?
- What do we need to do today to prevent such life-threatening hazards in the future, while keeping the benefits?
- When big data on entire populations are readily available, aspects of integrity and security become critical challenges to manage.

A supersmart society is a society where the various needs of society are finely differentiated and met by providing the necessary products and services in the required amounts to the people who need them when they need them, and in which all people can receive high-quality services and live a comfortable, vigorous life that makes allowances for their various differences such as age, sex, region, or language. It is called this to indicate a new society created by transformations led by scientific and technological innovation, following hunting society, agrarian society, industrial society, and information society (Harayama, 2017). In addition, society values are often attributes and not easy to measure; however, in many quality improvement projects, measurable quantities are often used. This is even more pronounced in the smart society described above, with the inclusion of big data and AI. The fact is that most human values, such as integrity, openness, trust, and transparency are not easily translated into quantities. That makes quality improvement projects within the context of Society 5.0 value-added where they are targeted for improvement.

## Industry 5.0 and Society 5.0

Industry 5.0 has three main components: human-centric collaboration, resilience and social responsibilities, and sustainability. It is a response to the need for Industry 4.0 to provide more human inclusion and environmental sustainability, complementing its predecessor, Industry 4.0. It does not, in fact, address social or societal considerations. The rise of Society 5.0 addresses this deficiency and presented a framework for a "supersmart society."

Society 5.0 (supersmart society) is a new guiding principle for innovation. It promotes convergence between cyberspace and physical space enabling AI-based on big data and robots to perform or support as an agent the work and adjustments that humans have done up to now (Fukuyama, 2018). As stated in Cabinet Office of Japan:

> Society 5.0 takes systemic approach but focusing on human beings with the aim to involve a wide variety of actors that in the past have only participated in non-visible ways (e.g. women and young people). It is a space for accommodating various bottom-up ideas. Society 5.0 calls for "systemization" of services and projects, more advanced systems, and coordination between multiple systems – thus aiming to serve as a Smart Bridge between the techno-centric and human-centric perspectives. Society 5.0 states that companies, universities, and others who

are responsible for innovation systems should strengthen their cooperation, breaking down organizational walls and promoting open innovation.

What should be strengthened is that Society 5.0 considers social capital its key asset, and it promotes globally targeted open innovation with human-centric priorities. In this concept, every citizen representing each generation has a role to play in the innovation process. The supersmart society is built upon delivering concrete, targeted and personalized, just-in-time solutions to the people, to provide a healthy and safe environment, and to promote well-being. It is still a vision, directive, or goal and not the reality. Yet, it opens a new perspective to understanding and using the technological advancement and digital transformation for the benefit of society (Carayannis and Morawska-Jancelewicz, 2022).

## Society 5.0 and Quality of Life

Society 5.0 aims to create a harmonious integration of advanced technology and human needs, driving sustainable development and improving quality of life. Slavko Arsovski (2019) thought of Society 5.0 as a product of four main characteristics: optimizing society with the integration of cyberspace and physical space, using data as a new resource, realizing an environment with change and constraints, and solving complex social and living problems. He also stated that all these characteristics are related to Quality of Life (QoL).

We can also argue that Society 5.0 has more characteristics that make the argument for its effectiveness even more realistic. It addresses the enhancement of human well-being and creating value for individuals, It provides a near-perfect integration of the physical and digital worlds; employs extensively digital tools, such as AI, IoT, big data, robotics, and automation; presents serious consideration of sustainability; and presents innovation as a vehicle to improve services in healthcare, education, and manufacturing.

The Society 5.0 concept promises enhanced quality of life, sustainability, economic growth, improved efficiency and inclusion; it also presents some challenges of accommodating the resistance to change, privacy, security, and ethical concerns.

## LOOKING TO THE FUTURE

Significant leaps forward in data and digital capabilities are surely changing the way the world moves, with an impact that will ripple through all sectors of business. This is the norm when revolutionary technologies are brought to the forefront and creative thinkers choose to push the limits of what can be done. Many aspects of the future around Quality 4.0 and beyond to Quality 5.0 will be incredibly challenging to predict, while providing unprecedented results. One thing is certain, however; during all previous industrial revolutions, technology eventually won out. There are no businesses today that do not use electricity or the machinery that became viable with steam. Programmable logic and electromechanical automation have been a mainstay in manufacturing for decades. The fourth industrial revolution will bring change as well, and businesses can embrace the change or choose to resist it. Embracing change, committing to navigating it effectively with growth, and executing a strong digital strategy are the steps you can take to achieve success in this new, data-driven world. Deploying Quality 4.0 is not only a path forward to leverage data as an advantage in the future, but a commitment to continue success in an ever-changing world. It's a bold first step toward Quality 5.0 and beyond.

# References

Abollado, J. Rojo, E. Shehab, and P. Bamforth. 2017. "Challenges and Benefits of Digital Workflow Implementation in Aerospace Manufacturing Engineering." *Procedia CIRP* 60: 80–85. https://doi.org/10.1016/j.procir.2017.02.044.

Abuzayed, Yasmin. 2024. "Research Report: History of Quality." Personal communications, Orlando, Florida.

Adel, Amr. 2022. "Future of Industry 5.0 in Society: Human-Centric Solutions, Challenges and Prospective Research Areas." *Journal of Cloud Computing* 11 (2022): 40. https://doi.org/10.1186/s13677-022-00314-5.

American Society for Quality. 2025. "Total Quality Management: What is Total Quality Management?" https://asq.org/quality-resources/total-quality-management.

Antony, Jiju., Michael Sony, and Elizabeth Cudney. 2020. "Quality 4.0: Motivations and Challenges from a Pilot Survey in European Firms." The Future Factory. https://www.thefuturefactory.com/blog/53.

Antony, Jiju, Olivia McDermott, Michael Sony, Aidan Toner, Shreeranga Bhat, Elizabeth A. Cudney, and Mehran Doulatabadi. 2022. "Benefits, Challenges, Critical Success Factors and Motivations of Quality 4.0–A Qualitative Global Study." *Total Quality Management & Business Excellence* 34 (7–8): 827–46. doi:10.1080/14783363.2022.2113737.

Antony, Jiju, Michael Sony, Sandra Furterer, Olivia McDermott, and Matthew Pepper. 2021. "Quality 4.0 and Its Impact on Organizational Performance: An Integrative Viewpoint." *The TQM Journal* 34 (6): 2069–2084. https://doi.org/10.1108/TQM-08-2021-0242.

Antony, Jiju, Olivia McDermott, and Michael Sony. 2022. "Quality 4.0 Conceptualisation and Theoretical Understanding: A Global Exploratory Qualitative Study." *The TQM Journal* 34 (5): 1169–1188. https://doi.org/10.1108/TQM-07-2021-0215.

Antony, Jiju, Michael Sony, Olivia McDermott, Raja Jayaraman, and David Flynn. 2023. "An Exploration of Organizational Readiness Factors for Quality 4.0: An Intercontinental Study and Future Research Directions." *International Journal of Quality & Reliability Management* 40 (2): 582–606.

Arsovski, Slavko. 2023. "Quality 5.0: From Challenges to Reality." *Journal of Innovations in Business and Industry* 1 (1): 13–21. doi: 10.61552/JIBI.2023.01.002.

Arsovski, Slavko. 2019. "Quality of Life and Society 5.0." *Quality Festival 2019: 13th International Quality Conference*, May 29-June 1, 2019. 1 (2): 775–780.

Arsovski, Slavko. 2019. "Social-Oriented Quality: From Quality 4.0 Towards Quality 5.0." *Quality Festival 2019: 13th International Quality Conference*, May 29-June 1, 2019. 1 (2): 397–404.

Babatunde, Oluwayomi Kayode. 2020. "Mapping the Implications and Competencies for Industry 4.0 to Hard and Soft Total Quality Management." *The TQM Journal* 33 (4): 896–914. doi:10.1108/TQM-07-2020-0158.

Balouei Jamkhaneh, Hadi, Arash Shahin, Sahar Valipour Parkouhi, and Reza Shahin. 2022. "The New Concept of Quality in the Digital Era: A Human Resource Empowerment Perspective." *TQM Journal* 34: 125-144. doi:10.1108/TQM-01-2021-0030.

Banks, Jerry. *Principles of Quality Control.* John Wiley & Sons Inc, 1989.

Bergman, Bo, Ingela Bäckström, Rickard Garvage, and Bengt Klefsjo. *Quality: From Customer Needs to Customer Satisfaction*. Lund, Sweden: Studentlitteratur, 2022.

Butt, Javaid. 2020. "A Conceptual Framework to Support Digital Transformation in Manufacturing Using an Integrated Business Process Management Approach." *Designs* 4 (3): 17. https://doi.org/10.3390/designs4030017.

Calp, M. Hanefi, and Resul Bütüner. 2022. "Chapter 7–Society 5.0: Effective Technology for a Smart Society." In *Artificial Intelligence and Industry 4.0*, Academic Press, 2022: 175–194. https://doi.org/10.1016/B978-0-323-88468-6.00006-1.

Carayannis, Elias G. and Joanna Morawska-Jancelewicz. 2022. "The Futures of Europe: Society 5.0 and Industry 5.0 as Driving Forces of Future Universities." *Journal of the Knowledge Economy* 13 (4): 3445–3471.

Cervone, H. Frank. 2016. "Organizational Considerations Initiating a Big Data and Analytics Implementation." *Digital Library Perspectives* 32 (3): 137–141. doi:10.1108/DLP-05-2016-0013.

Chiarini, Andrea, and Maneesh Kumar. 2021. "What Is Quality 4.0? An Exploratory Sequential Mixed Methods Study of Italian Manufacturing Companies." *International Journal of Production Research* 60 (16): 4890–4910. doi:10.1080/00207543.2021.1942285.

Collins, Jim. *Good to Great: Why Some Companies Make the Leap and Others Don't.* HarperCollins, 2001.

Cudney, Elizabeth A. and Elizabeth M. Keim. 2017. "The Changing Role of Quality in the Future: Required Competencies for Quality Professionals to Succeed." *Journal of Quality and Participation* 39 (4): 4–11.

Dinana, Hesham. 2020. "Society 5.0 and The Future of Economics." *The Cairo Review of Global Affairs*, Spring. Retrieved from: https://www.thecairoreview.com/essays/society-5-0-and-the-future-economies/ on June 3, 2024.

Directorate General for Research and Innovation. 2022. "Industry 5.0, a Transformative Vision for Europe," European Commission, January 10, 2022.

Ellis, Whitney. 2024. "Research Report: Quality 1.0 through Quality 4.0." Personal communications, Orlando, Florida.

Elshennawy, Ahmad. 2019. "Quality 4.0: Challenges for Implementing Quality Management Systems of the Future." Invited keynote presentation at the 4th North American IEOM Conference, Toronto, Canada, October 23–25, 2019.

Evans, James R. and William M. Lindsay. *Managing for Quality and Performance Excellence*, 11th ed. Cengage Learning, 2020.

Farooq, Abdur Rahman. "Evolution of Quality Management: From Quality 4.0 to Quality 5.0." LinkedIn. Published Apr 17, 2024. Downloaded from: https://www.linkedin.com/pulse/evolution-quality-management-from-40-50-abdur-rahman-farooq-pqcuc/. Retrieved June 3, 2024.

Fincham, Jack E. 2008. "Response Rates and Responsiveness for Surveys, Standards, and the Journal." *American Journal of Pharmaceutical Education* 72 (2): 43. doi:10.5688/aj720243.

Ford, Henry and Samuel Crowther. *Today and Tomorrow*. Doubleday, Page & Company, 1926.

Frick, Jan and Piotr Grudowski. 2023. "Quality 5.0: A Paradigm Shift Towards Proactive Quality Control in Industry 5.0." *Asia-Pacific Journal of Business Administration*. 14 (2): 51–56. https://doi.org/10.5430/ijba.v14n2p51.

Fukuyama, Mayumi. 2018. "Society 5.0: Aiming for a New Human-Centered Society." *Japan SPOTLIGHT*, July/August: 47–50.

Goel, Ruchi and Pooja Gupta. 2020. "Robotics and Industry 4.0." In *A Roadmap to Industry 4.0: Smart Production, Sharp Business and Sustainable Development*. Anand Nayyar and Akshi Kumar, eds. *Advances in Science, Technology & Innovation series*. Springer, Cham. https://doi.org/10.1007/978-3-030-14544-6_9.

Gunasekaran, Angappa, Nachiappan Subramanian, and Wai Ting Eric Ngai. 2019. "Quality Management in the 21st Century Enterprises: Research Pathway Towards Industry 4.0." *International Journal of Production Economics* 207: 125–129.

Halpern, Nigel, Deodat Mwesiumo, Pere Suau-Sanchez, Thomas Budd, and Svein Bråthen, S. 2021. "Ready for Digital Transformation? The Effect of Organisational Readiness, Innovation, Airport Size and Ownership on Digital Change at Airports." *Journal of Air Transport Management* 90 (1): 101949–101960.

Hamid, Syaiful Rizal, Saifuddin Isa, Boon Cheong Chew, and Abdullah Altun. 2019. "Quality Management Evolution from the Past to Present: Challenges for Tomorrow." *Organizacija* 52 (3): 157–186. https://doi.org/10.2478/orga-2019-0011.

Harayama, Yuko. 2017. "Society 5.0: Aiming for a New Human-Centered Society – Japan's Science and Technology Policies for Addressing Global Social Challenges. *Hitachi Review* 66 (6): 554–555, 558–559.

Harrington, H. James and Frank Voehl. 2019. Personal communications.

Hattinger, Monika and Kostas Stylidis. 2023. "Transforming Quality 4.0 towards Resilient Operator 5.0 Needs." *Procedia CIRP* 120: 1600–1605. https://doi.org/10.1016/j.procir.2023.12.002.

Harayama, Yuko. 2017. "Society 5.0: Aiming for a New Human-Centered Society." Collaborative Creation Through Global R&D Open Innovation for Creating the Future. *Hitachi Review* 66 (6): 558–559.

Havle, Celal Alpay and Caglar Ucler. 2018. "Enablers for Industry 4.0." IEEE International Symposium on Multidisciplinary Studies and Innovative Technologies (ISMSIT), Ankara, Turkey.

Hizam-Hanafiah, Mohd, Mansoor Ahmed Soomro, and Nor Liza Abdullah. 2020. "Industry 4.0 Readiness Models: A Systematic Literature Review of Model Dimensions." *Information* 11 (7): 364-371. https://doi.org/10.3390/info11070364.

Infinity QS. 2017. "Inside the Next Factory of the Future: Industry 5.0." https://www.infinityqs.com/blog/april-2017/inside-the-next-factoryof-the-future-industry-5.

Jacob, Dan. 2017. "Quality 4.0: Fresh Thinking for Quality in the Digital Era." *Quality Digest*. Retrieved from: https://www.qualitydigest.com/inside/management-article/quality-40-072417.html.

Jacob, Dan and LNS Research. 2017. *The Quality 4.0 Impact and Strategy Handbook*. LNS Research. Retrieved from: https://generisgp.com/2018/02/15/the-quality-4-0-impact-and-strategy-handbook/.

Javaid, Mohd, Haleem Abid Haleem, Ravi Pratap Singh, and Rajiv Suman. 2021. "Significance of Quality 4.0 Towards Comprehensive Enhancement in Manufacturing Sector." *Sensors International* 2: 100–109. https://doi.org/10.1016/j.sintl.2021.100109.

Jones, Thomas L. 2024. "Qualifying Quality 4.0." *The Quality Management Forum* 49: 4.

Jones, Thomas L. 2023. "A Conceptual Model for Quality 4.0 Deployment in U.S. Based Manufacturing Firms." Doctoral dissertation, University of Central Florida. https://stars.library.ucf.edu/etd2020/1584.

Juran Institute. 2019. "Quality 4.0: The Future of Quality." https://www.juran.com/blog/quality-4-0-the-future-of-quality/. Retrieved May 29, 2024.

Kannan, K. Sumitha P. N. and Alaa Garad. 2020. "Competencies of Quality Professionals in the Era of Industry 4.0: A Case Study of Electronics Manufacturer from Malaysia." *International Journal of Quality & Reliability Management* 38 (3): 839–871. https://doi.org/10.1108/IJQRM-04-2019-0124.

Khan, Lina. 2023. "A Framework of Critical Success Factors for Implementation of Industry 4.0 in Aerospace and Defense Industries." Doctoral dissertation, University of Central Florida. https://stars.library.ucf.edu/etd2020/1589.

Knox, Tyler J. 2020. "U.S. Military Innovation in the 21st Century: The Era of the "Spin-On." Doctoral dissertation, University of Pennsylvania, https://repository.upenn.edu/handle/20.500.14332/49239.

Küpper, Daniel, Claudio Knizek, Dave Ryeson, and Jan Nöcker. 2019. "Quality 4.0 Takes More Than Technology." Boston Consulting Group. https://www.bcg.com/publications/2019/quality-4.0-takes-more-than-technology.

Lawless, William F., Ranjeey Mittu, and Donald A. Sofge. 2020. "Chapter 1 – Introduction: Artificial Intelligence (AI), Autonomous Machines, and Constructing Context: User Interventions, Social Awareness, and Interdependence." In *Human-Machine Shared Contexts* (1–22). Academic Press, 2020. https://doi.org/10.1016/B978-0-12-820543-3.00001-8.

Lewis, Dustin. A., Gabriella Blum, and Naz K. Modirzadeh. 2016. "War-Algorithm Accountability." Harvard Law School Program on International Law and Armed Conflict Research Briefing. SSRN 2832734.

Liu, Hu-Chen, Xiuzhu Gu, and Miying Yang. 2023. "From Total Quality Management to Quality 4.0: A Systematic Literature Review and

Future Research Agenda." *Frontiers of Engineering Management* 10 (2): 191–205. https://doi.org/10.1007/s42524-022-0243-z.

Loy, Adrian Chun Minh, Bridgid Lai Fui Chin, and Revathy Sankaran. 2021. "Industrial Revolution 1.0 and 2.0." In *The Prospect of Industry 5.0 in Biomanufacturing* (1st ed): 1–14. CRC Press. https://doi.org/10.1201/9781003080671-1-1.

Martin, Jason, Mattias Elg, and Ida Gremyr. 2020. "The Many Meanings of Quality: Towards a Definition in Support of Sustainable Operations." *Total Quality Management & Business Excellence* 36 (3–4): 185–98. doi:10.1080/14783363.2020.1844564.

Masood, Tariq and Johannes Egger. 2019. "Augmented Reality in Support of Industry 4.0 – Implementation Challenges and Success Factors." *Robotics and Computer-Integrated Manufacturing* 58: 181–195. https://doi.org/10.1016/j.rcim.2019.02.003.

Massaro, Alessandro, Ivano Manfredonia, Angelo Galiano, and Nicola Contuzzi. 2019. "Inline Image Vision Technique for Tires Industry 4.0: Quality and Defect Monitoring in Tires Assembly." *2019 II Workshop on Metrology for Industry 4.0 and IoT* (MetroInd4.0&IoT), Naples, Italy, June 4-6 2019: 54–57. doi:10.1109/METROI4.2019.8792911.

Massaro, Alessandro, Ivano Manfredonia, Angelo Galiano, Leonard Pellicani, and Vitangelo Birardi. 2019. "Sensing and Quality Monitoring Facilities Designed for Pasta Industry Including Traceability, Image Vision and Predictive Maintenance." *2019 II Workshop on Metrology for Industry 4.0 and IoT (MetroInd4.0&IoT)*, Naples, Italy, June 4–6, 2019: 68–72. doi:10.1109/METROI4.2019.8792912.

Montgomery, Douglas C. *Statistical Quality Control* (Vol. 7). New York: Wiley, 2009.

Neumann, W. Patrick, Sven Winkelhaus, Eric H. Grosse, and Christoph H. Glock. 2021. "Industry 4.0 and the Human Factor – A Systems Framework and Analysis Methodology for Successful Development." *International Journal of Production Economics* 233: 107992.

Novak, William E. 2021. "Artificial Intelligence (AI) and Machine Learning (ML) Acquisition and Policy Implications." Software Engineering Institute, Carnegie Mellon University.

Ohno, Taiichi. 1988. *Toyota Production System: Beyond Large-Scale Production* (1st ed.). Productivity Press. https://doi.org/10.4324/9780429273018.

Olaitan, J. and Elshennawy. 2024. "Human-Centric Quality Management in Industry 5.0: Bridging People and Technology in Digital Manufacturing." *2024 International Conference on Industry, Engineering, and Management Systems (IEMS 25)*, Orlando, Florida, March 3–5, 2024.

Papke, Barry L., Gan Wang, Ron Kratzke, and Chris Schreiber. 2020. "Implementing MBSE – An Enterprise Approach to an Enterprise Problem." *INCOSE International Symposium*, 30 (1): 1539–1556. https://doi.org/10.1002/j.2334-5837.2020.00803.

Permana, Adi and Humiras Purba. 2021. "A Systematic Literature Review of Total Quality Management (TQM) Implementation in the Organization. *International Journal of Production Management and Engineering* 9 (1): 25–36. doi:9.10.4995/ijpme.2021.13765.

Petersen, Marcel. 2024. "Total Quality Management – More than Just Quality Management." GBTEC, www.gbtec.com/resources/total-qualitymanagement/.

Ponraj, Ram Prakash, Vijay Ravindran, Satheesh Ragunathan, K. Swaminathan, and Titus Sigamani. 2024. "Chapter 3: Society 5.0 and Explainable Artificial Intelligence – Implications." From *XAI Based Intelligent Systems for Society 5.0*. Elsevier, 2024: 61–101. https://doi.org/10.1016/B978-0-323-95315-3.00009-7.

Pozzi, Rossella, Tommaso Rossi, and Raffaele Secchi. 2021. "Industry 4.0 Technologies: Critical Success Factors for Implementation and Improvements in Manufacturing Companies." *Production Planning & Control* 34 (2): 139–58. doi:10.1080/09537287.2021.1891481.

Quality World. 2023. "Interview: Lars Sorqvist – Advancing Quality for Humanity." *Quality World* magazine, Spring 2023: 20–25.

Radziwill, N. M. 2018. "What is Quality 4.0? Quality and Innovation – Performance Excellence and Data Science for Digital Transformation in Industry 4.0." QualityandInnovation.com. Retrieved from: https://qualityandinnovation.com/2018/01/02/what-is-quality-4-0/.

Radziwill, N. M. *Connected, Intelligent, Automated: The Definitive Guide to Digital Transformation and Quality 4.0*. Milwaukee: ASQ Quality Press, 2020.

Reis, Marco S. and Pedro M. Saraiva. 2022. "Data-Driven Process System Engineering: Contributions to Its Consolidation Following the Path Laid Down by George Stephanopoulos." *Computers and Chemical Engineering* 159: 1–15. https://doi.org/10.1016/j.compchemeng.2022.107675.

Sader, Sami, Istvan Husti, and Miklos Daroczi. 2021. "A Review of Quality 4.0: Definitions, Features, Technologies, Applications, and Challenges." *Total Quality Management & Business Excellence* 33 (9–10): 1164–82. doi:10.1080/14783363.2021.1944082.

Saihi, Afef, Mahmoud Awad, and Mohamed Ben-Daya. 2023. "Quality 4.0: Leveraging Industry 4.0 Technologies to Improve Quality Management Practices – A Systematic Review." *International Journal of Quality and Reliability Management* 40 (2): 628–650. https://doi.org/10.1108/IJQRM-09-2021-0305.

Salimbeni, Sergio, Andrés Redchuk, and Hilda Rousserie. 2023. "Quality 4.0: Technologies and Readiness Factors in the Entire Value Flow Life Cycle." *Production & Manufacturing Research* 11 (1). doi:10.1080/21693277.2023.2238797.

Santos, Gilberto, Jose Carlos Sá, Maria João Félix, Luís Barreto, Filipe Carvalho, Manuel Doiro, Kristína Zgodavová, and Miladin Stefanović. 2021. "New Needed Quality Management Skills for Quality Managers 4.0." *Sustainability* 13 (11): 6149. https://doi.org/10.3390/su13116149.

Saraiva, Pedro. 2023. "Quality 5.0 in Turbulent Times: How to Think About and Apply Quality in Today's World of Societal Challenges." *Quality Progress* 56 (10): 18–25.

Sony, Michael and Subhash Naik. 2019. "Key Ingredients for Evaluating Industry 4.0 Readiness for Organizations: A Literature Review." *Benchmarking: An International Journal* 27 (7): 2213–2232. https://doi.org/10.1108/BIJ-09-2018–0284.

Sony, Michael and Subhash Naik. 2020. "Critical Factors for the Successful Implementation of Industry 4.0: A Review and Future Research Direction." *Production Planning and Control* 31 (10): 799–815. doi:10.1080/09537287.2019.1691278.

Sony, Michael., Jiju Antony, and Jacqueline Ann Douglas. 2020. "Essential Ingredients for the Implementation of Quality 4.0: A Narrative Review of Literature and Future Directions for Research." *The TQM Journal* 32 (4): 779–793. https://doi.org/10.1108/TQM-12-2019-0275.

Sony, Michael, Jiju Antony, Jacqueline Ann Douglas, and Olivia McDermott. 2021. "Motivations, Barriers and Readiness Factors for Quality 4.0 Implementation: An Exploratory Study." *The TQM Journal*, 33 (6): 1502–1515. https://doi.org/10.1108/TQM-11-2020-0272.

Souza, Fabiane Florencio de, Alana Corsi, Regina Negri Pagani, Giles Balbinotti, and João Luiz Kovaleski. 2022. "Total Quality Management 4.0: Adapting Quality Management to Industry 4.0." *The*

*TQM Journal* 34 (4): 749–769. https://doi.org/10.1108/TQM-10-2020-0238.

Stockwell, Scott. 2017. "A Framework for Industry 4.0, 9 minute read." https://www.ibm.com/blogs/internet-of-things/industry-4-0-industrial-framework/. Retrieved October 17, 2019.

Swedish Institute for Quality. "Aiming for Sustainable Success." www.siq.se.

Viswakarma, Aryan. 2021. "What is Quality 4.0? What is impact of Quality 4.0 to Industry?" Quality HUB India. Retrieved from https://qualityhubindia.com/what-is-quality-4-0/.

Wang, Gan. 2020. "Implementing a Model-Based, Digital Engineering Enterprise for a Defense Systems Integrator – an Ongoing Journey." *INCOSE International Symposium* 30 (1): 783–798. https://doi.org/10.1002/j.2334-5837.2020.00755.

Xu, Xiaoya and Qingsong Hua. 2017. "Industrial Big Data Analysis in Smart Factory: Current Status and Research Strategies." *IEEE Access* 5: 17543–17551. doi:10.1109/ACCESS.2017.2741105.

Yin, Yong, Kathryn E. Stecke, and Dongni Li. 2017. "The Evolution of Production Systems from Industry 2.0 through Industry 4.0." *International Journal of Production Research* 56 (1–2): 848–61. doi:10.1080/00207543.2017.1403664.

Zonnenshain, Avigdor, and Ron S. Kenett. 2020. "Quality 4.0 – The Challenging Future of Quality Engineering." *Quality Engineering* 32 (4): 614–26. doi:10.1080/08982112.2019.1706744.

# About the Authors

**Thomas Jones** has bachelor's, master's, and doctoral degrees in industrial engineering with specializations in Quality Systems and digital transformation. Currently, he works as an Industrial Engineer in Orlando, Florida; his experience includes the Aerospace/Defense and Automotive industries. He has extensive experience in new product development and introduction, operational excellence including Lean Six Sigma, and stabilizing processes. Thomas is a Senior Member of ASQ with the following certifications: CRE, CQE, CSSBB, CMQ/OE, and CQA. Through his work with ASQ, he is also a contributor to the fourth edition of *The ASQ Certified Reliability Engineer Handbook.*

**Lina Khan** has a bachelor's degree in mechanical engineering, a master's degree in engineering management, and a doctoral degree in industrial engineering. Currently, she works in the Aerospace and Defense Industry as a Model-Based Systems Engineer (MBSE) in Cape Canaveral, Florida. Lina focuses on solutions, automation, digital transformation, process development, and improvement efforts. She has held various critical roles including engineering, program management, business development, and operations.

**Ahmad Elshennawy** is a professor of industrial engineering and executive director of the University of Central Florida (UCF) Quality Institute. He has bachelor's and master's degrees in production engineering from Alexandria University (Egypt) and a master's and a doctoral degree in industrial engineering from Penn State University. He is a Certified Quality Engineer (CQE), Certified Reliability Engineer (CRE), and CLSSMB and Academician at the International Academy for Quality (IAQ). He has written multiple books published by Quality Press.

www.ingramcontent.com/pod-product-compliance
Lightning Source LLC
LaVergne TN
LVHW022209200926
848481LV00006B/1903
* 9 7 8 1 6 3 6 9 4 1 8 2 0 *